The Jay,
the Beech
and the
Limpetshell

The Jay,
the Beech
and the
Limpetshell

*Finding Wild Things
With My Kids*

RICHARD SMYTH

ICON

Published in the UK and USA in 2023 by
Icon Books Ltd, Omnibus Business Centre,
39–41 North Road, London N7 9DP
email: info@iconbooks.com
www.iconbooks.com

Sold in the UK, Europe and Asia
by Faber & Faber Ltd, Bloomsbury House,
74–77 Great Russell Street,
London WC1B 3DA or their agents

Distributed in the UK, Europe and Asia
by Grantham Book Services, Trent Road,
Grantham NG31 7XQ

Distributed in the USA
by Publishers Group West,
1700 Fourth Street, Berkeley, CA 94710

Distributed in Canada by Publishers Group Canada,
76 Stafford Street, Unit 300
Toronto, Ontario M6J 2S1

Distributed in Australia and New Zealand by
Allen & Unwin Pty Ltd, PO Box 8500,
83 Alexander Street, Crows Nest, NSW 2065

Distributed in South Africa by
Jonathan Ball, Office B4, The District,
41 Sir Lowry Road, Woodstock 7925

Distributed in India by Penguin Books India,
7th Floor, Infinity Tower – C, DLF Cyber City,
Gurgaon 122002, Haryana

ISBN: 978-178578-802-4

Text copyright © 2023 Richard Smyth

Text designed and set in Monotype Dante by Tetragon, London
Printed and bound in the UK.

For my family, *up* and *down*

CONTENTS

AUTHOR'S NOTE

This book was not written in the order in which you're probably reading it. However, my children, who appear at times herein, chose unhelpfully to age as I wrote, and in a conventional, linear fashion. You'll see as you read on that their ages jump about, with the bigger one being three here and two there, the smaller one being at one point a toddler and at some later point a baby. I'm sorry if this is confusing. There was very little I could do about it. I will say one thing. Genevieve, if you're reading this, don't worry – because I know it bothers you, sometimes – you will always be the big sister, and Daniel will always be the little brother, no matter how big he gets.

Prologue

HOBBES: Whatcha doin'?

CALVIN: Looking for frogs.

HOBBES: How come?

CALVIN: I must obey the inscrutable exhortations of my soul.

HOBBES: Ah, but of course.

CALVIN: My mandate also includes weird bugs.

BILL WATTERSON,
Calvin and Hobbes, 13 March 1995

My daughter's pockets are empty. She's three, and she doesn't want to put things in her pockets – what use are they there? The things we find, shells, sticks, feathers, pebbles, leaves, fragments of ice, she wants to carry them (handle them, taste them, break them) or she wants Daddy or Mummy to look after them. There's the long cone of a pine tree from Northumberland in the pocket of our car's offside front door. There's a daisy wilting in an espresso cup of water on the kitchen windowsill. There's a small stack of sticks in the hall ('Have you got a dog?' the postman asks). My own coat pockets fill up with snail shells and alder cones. She doesn't know a lot about any of these things – she's three, give her a break – but I know she'll learn.

Then there's our son. He was born at the beginning of the Covid-19 pandemic. He knows what crows and blackbirds are, that ducks go 'quack' and frogs go 'ribbit'. He's watched a lot of David Attenborough (we binge my old boxsets in the early mornings). He can snap his jaws like a crocodile and nearly say 'elephant' (*efefe*). He's been walked and wheeled and carried a long way along footpaths, towpaths, park paths, river paths. He'll learn, too.

I wonder what sort of world they'll learn about. It won't be the same world I learned about when I was a kid. They're already used to the cork-on-glass squealing of rose-ringed parakeets in the local park. They'll soon come to know the silhouettes of red kites. My daughter knows that the big bird that sometimes comes to sit on the mill chimney by our house is a peregrine. None of these birds would have been here when I was a kid.

My children's world will be warmer, wetter, more populous, more depleted (less greenery, fewer animals, fewer *sorts* of animals). Will it be worse? I don't know. I know – I *think* I know – that however bad it is, they'll make it better. I know that from some perspectives the decision to bring more humans into a world already replete, overstuffed, with humanity is a contentious one; that children are quite frankly problematic, and not only in the eating-the-soap and shaving-the-cat sense. The short version of my argument on this is that humans are all right really. In the main, on balance, at bottom, when all is said and done, they're all right. We're all right. Let's carry on and see what we can do.

So here we are, my wife and I, with these children of ours. Here are woodpigeons on the roof, magpies in the road, owl calls in the night, finch songs in next door's laburnum, snails

in the yard, frogs in the lightwell. Here are songs about bun-
nies, toy lions, bath whales, fox socks, penguin T-shirts. Here
are YouTube videos where enthusiastic young Americans
explain about awesome sharks and cool crocs and amazing bees
and incredible octopuses. Here are Maddie Moate and Steve
Backshall on the television; here are JoJo and Gran Gran going
birdwatching or pond-dipping; here are *Octonauts* and *Peter
Rabbit*, *Ferne and Rory's Vet Tales* and *Down on the Farm*; here's
The Jungle Book on DVD and *Ladybug* on Amazon Prime. Here
are books, dozens, thousands of books; *Sharing a Shell*, *Handa's
Surprise*, *The Last Polar Bears*, *The Bad-Tempered Ladybird*, *There's
an Ouch in My Pouch!*, *Who's Afraid of the Dark?*, *Sleepy Places*,
Snuggle Up Sleepy Ones, *Hunwick's Egg*.

Here, in short, is a child's world full of wild things, and
here's us trying to figure out how to make sure they grow
up to know it, and love it, and look after it, and, who knows,
maybe even save it. Where do you start?

I'll start here: with the stuff we've picked up, the things
we've collected. The medieval mystic Julian of Norwich saw
the whole world, *all that is made*, in a hazelnut. This shell, these
bones, this lost feather, this beech leaf: I think there are worlds
to be seen in these things, too.

Charles Waterton, a squire of the early 19th century who
lived in Walton, near Wakefield, just down the road from
where I grew up, was one of the great early naturalists. His
biographer, Julia Blackburn, has written that throughout his
life 'he maintained a tactile approach to the external world.

He wanted to taste it, to roll in it, to get closer and closer still to everything which invited his curiosity.' At the age of three he swallowed the egg of a meadow lark. As an adult he tried to eat part of a swallow's nest, to determine its flavour ('I have chewed a piece for a quarter of an hour but found it absolutely tasteless'). These days, Waterton is mostly remembered – if he's remembered at all – as an eccentric. He was certainly quixotic. He was a bold and kind-hearted conservationist; he was also, in very many ways, childlike.

Our urge to touch and grab is one that comes to us – takes hold of us – very early, and one that we very quickly learn to deny. *Look, don't touch*, is the standard injunction (the second warning, when I was small, usually took the form *you look with your eyes, not with your hands*). Confronted with something interesting, something that looks interesting and looks like it might *feel* interesting, or, even better, looks like we might be able to seize it and do interesting things *with* it, little hands steal out almost unbidden, almost more like antennae, 'feelers', mobile sense organs, than hands. It comes to us early and it never goes away. Think of, say, an art gallery: don't your hands just itch to feel the cold curve of sculpted marble, the textured oils of an impasto painting; don't you want to heft that rare ceramicware in your hands, just for a moment, just to see if it's heavier than you thought, or lighter than you thought, to see where its balance lies, get a sense of its gravity?

I can't not pick up a toad if I find one. A toad, discovered under a stone or a rotting log, is, somehow, a very *there* thing. It's compact and complete and still and it will as a rule sit fairly stoically in the cupped palm of your hand for a minute or so before it decides it has somewhere else to be. I don't know what exactly I'm looking to learn by picking up the toad. I'm not going to advance herpetological knowledge in any meaningful way; I just somehow feel that until I've held it I haven't really seen it.

You look with your eyes, not with your hands. Perhaps that's not quite as true as it seems. I've no desire to smell or taste the toad, but touching is different, holding is different. These things somehow seem closer to *knowing*.

They also come with risks attached. The European toad, for instance, carries, in the bulging parotoid glands behind its eyes, a substance called bufotoxin, which is related to digitalis, the foxglove toxin. It isn't dramatically harmful to humans, but it's unpleasant. That's how most of the living biohazards here in the United Kingdom are best described: unpleasant. A nettle sting is unpleasant. A nip from a red ant is unpleasant. We get rashes and hives, sore spots, prickles, scratches, swellings. A wasp sting hurts for a bit. A hornet sting hurts for a lot. Step on the spines of a lesser weever fish while paddling in the sea and you'll know about it (numbness if you're lucky, excruciating pain if you're not). An adder bite is about as serious as it gets here – but there have been only fourteen recorded human deaths from adder bites since 1876, and none at all in my lifetime (it may or may not be a comfort to know that you are *far* more likely to die from an allergic reaction to a wasp or bee sting: around ten people a year die in the United Kingdom this way).

I remember my first bee sting quite vividly. I remember that I had it coming. I was quite small. I don't remember exactly what I was doing or trying to do with the honeybee, but it was probably not in the honeybee's best interests. I remember the onset of a sudden intense curiosity, and then a sudden panic as the bee latched on to my thumb tip, all six legs closing tight like the bucket of a grab crane, and its abdomen lifted, and the stinger went *boink* straight into the fleshy ball of my thumb, and I yelled the house down. I had it coming, like I said.

I've often read, usually in books by older naturalists, from a more careless time, about how stinging insects like bees and wasps are quite easily picked up from behind with a firm finger and thumb about the abdomen. I'm very sure it is easy, but I still can't do it. Try it. I just tried it with a bee on our campanula out the back. As your finger and thumb creep closer, the humming of the bee's wingbeats rises warningly in pitch (or does it just *seem* that it does?) – then, if you're me, something inside you, something that feels quite basic, quite deep in the system, says *nope*, and you take your hand away. I'm always amazed by how hard this is for me to overcome. It's only a bee, man! Pull yourself together! But there it is, still: *nope*.

'Put it on my hand,' Genevieve said yesterday when I found a millipede in the front yard. She said it without hesitation. Clearly this is what millipedes are *for*. So I did. The millipede, a 'snake' millipede I think, an inch long and glossy coffee-brown, crawled smoothly and steadily up over her palm and across the heel of her hand and over the summit of the pisiform bone in her wrist, all its many little feet busy in a rolling shuffle, its many little footsteps all but imperceptible.

Living in the United Kingdom, we can allow ourselves to be pretty cavalier about this sort of thing. Some of our

8

millipedes release bad smells if they feel threatened, but that's about it.

In fact, there are no venomous millipedes anywhere in the world. Millipedes are the gentle vegetarians, the browsing herbivores, of the topsoil: humus cows, leaf mould sheep. Centipedes – fewer legs, more attitude – are a different proposition: they are hunters, and they do carry venom, but still, they can't do much damage to a human, even a three-year-old one. I'm not sure Genevieve would enjoy one scampering up her inner arm but it wouldn't *hurt*.

For all the manufactured bug panics that periodically infest our tabloid press – mutant fleas! killer wasps! four-inch crane-flies! (and this in a country where some hold out hope of one day reintroducing *wolves*) – there isn't very much to fear here. Of course, a wolf spider on the counterpane will give you pause. Of course, no one is fond of an unexpected earwig, especially if one drops without warning out of the nectarine you are cutting up for your breakfast and starts performing a wild dance of freedom on your plate (as happened last summer to my wife: it sounds like a lost chapter from *James and the Giant Peach*, but the event left us badly shaken and neither one of us will ever fully trust a nectarine again). But we do not have the katipō or the redback or the black widow spider, we do not have the Asian giant hornet, we do not have the deathstalker scorpion, we do not have the velvet ant *Dasymutilla occidentalis*, better known as the 'cow killer'.

If we did, we perhaps wouldn't be quite so gung-ho about putting millipedes on our children. Our relationships with our bugs, our insects, arachnids, arthropods and all the rest – and then, by extension, with the rest of the wild things among which we live – have been shaped in these mild British landscapes by the general geniality of most of our local species. We can, if we want, be pretty cosy with these creatures; we can, if we're sensible, and do what must be done about allergies and hygiene and so on, and teach good lessons about gentleness and care (*kind hands*, I yell, standing over my bug-hunting children like a rugby union referee over a ruck, *kind hands!*), try to raise our little naturalists to be hands-on.

There is an obvious caveat: don't let them eat anything. We can't be alone in having dealt with a variant on that old joke about the only thing worse than finding a worm in your apple. *What's worse than finding your toddler playing with a slug? Finding your toddler playing with half a slug.*

It would be wrong, anyway, to overplay this idea that only those of us raised in relatively harmless ecologies regard nature as a sort of open-access petting zoo. My guess is that everyone everywhere starts out that way; it's just that some people, in some places, have to learn pretty quickly that that's not always how this works. On Twitter the other day I saw an entomologist share a picture someone had posted of their cupped hand holding a pretty-looking red-and-black insect. *Someone just posted this photo on Reddit asking for an identification*, the entomologist wrote, *and just looking at it makes me wince.*

It's a cow killer – the velvet ant, actually a wingless wasp, but whatever you call it the possessor of one of the most painful stings in the insect world (its closest rival is the bullet ant, so called because its sting feels like you've been shot: I've consulted an authority, i.e. a guy on YouTube who lets himself get stung by stuff, and his exact words were *oh my gosh, d'oh my gosh, d'oh my god, it's really bad, oh my gosh, it is like, oooorrrggggg, it's really HOT*). My tip: don't hold one in your bare hand.

It was illuminating that even someone in the eastern United States – where cow killers are not rare, and there's no shortage of other small things that could cause you serious pain – wouldn't think twice about picking one up to say hello. But even more illuminating was one of the replies to the tweet. *Stung by 1 as a child*, wrote @MandytheFerret. *Was my fault. I had known what it was BEFORE messing with it, so I did it when my folks weren't at home. They had no sympathy*. This strikes at a fundamental truth about kids and bugs, and, therefore, about bugs and all of us. Do we *want* to be stung? No – not exactly. But we do want to *see*, to see in the richest sense, in the sense that goes some way beyond the collection of photons on a retina, the processing of images in the brain's occipital lobe. We want to see – and if getting stung is part of that package, then so be it. When we actually *get* stung (or bitten or mauled or chased or prickled or poisoned or infested or eaten) we may regret the bargain. But it's a bargain we'd strike again in a heartbeat.

On Genevieve's first – and so far only – trip to London, in the winter of 2018–19, we took her to see the animals. She was about four months old.

We saw walrus and brown bear, gorilla and painted dog, hellbender salamander and Qinling snub-nosed monkey. We saw a grey whale calf breaching the waves of Magdalena Bay and a spotted-tail quoll breaking cover to hunt birds in a New

South Wales rainforest. We saw an iguana peering from the eye socket of a rotting sealion carcass. We saw a Galápagos ground finch pecking blood from the living body of a Nazca booby.

Genevieve was in the forward-facing baby sling, and I held her literally nose-to-nose with each backlit photograph as we made our slow way around the exhibition. The colours of the images – ice white, blood red, tiger orange, the gold and baby-blue of the snub-nosed monkey – washed across her face. She was wide-eyed (she was never not wide-eyed, at four months, unless she was asleep). It all went in. I don't know much stuck.

It had become a Christmas tradition for us, for me and Catherine: each year, we'd visit my brother James and his wife Clare in London, and over the course of a December weekend we'd see a show, go for dinner, take in something festive in Hyde Park or at Kew, and – of course! – visit the Wildlife Photographer of the Year exhibition at the Natural History Museum in Kensington.

We haven't been since – first there was pregnancy and Danny, and then there was the pandemic – and I miss it.

I just had another look at the shortlisted pictures from that year. Some have stayed with me: a multitude of lost mayflies captured in streetlighting on a bridge over the Ebro by José Manuel Grandío; Ricardo Núñez Montero's haunting picture of a mother gorilla with the long-dead body of her child. Others I don't remember seeing at all. Again, I wonder which, if any, made a mark on my daughter – if only for their shapes and colours, or the contrasted brightness of an eye or a feather, or for something Mummy or Daddy or her uncle or auntie might have said ('Jesus!', 'Not keen on this one', 'Right place, right time, I suppose', 'Unbelievable', 'Bloody hell').

I've been trying to think about what single image made the greatest impression on me as a kid – did the most, I mean, to inspire or nourish my fascination for nature, for the wild, for living things. And I think the answer is a room full of dead things. Specifically, the room full of dead things curated and kept by Jean-Henri Fabre.

Fabre was a French entomologist and writer, noted for his close observations of the behaviour of insects and other creatures. He was born in 1823 and grew up poor in Aveyron in the rural south of France; when he was a boy, local people would see him crouched motionless over an anthill, a spider's nest, a weevil on a vine, and take him for an idiot. Toward the end of 1879, after leaving a career in teaching and settling near Orange, close to Avignon, Fabre published *Souvenirs Entomologiques*, his most celebrated work. On the profits, he acquired a property at Sérignan-du-Comtat, which he named 'Harmas'. It's now a museum that bears his name and houses his collections.

It's a little ironic, a little unfair, that when I think of Fabre I think of his collections, because he was one of the first naturalists – Charles Waterton was another – to place a proper emphasis on field observation, on real-world observations of living things. 'His writings,' Gerald Durrell wrote in *The Amateur Naturalist*, 'meant that you were suddenly transported out into the open air instead of, as with so many Victorian naturalists, into a museum.'

Those collections, though. In the photograph I remember – taken from a corner of the room, cut across by a shaft of Provençal sunlight through a high window – it's not always clear what all the things are. But to me, at seven or eight, they are, as Howard Carter whispered to Lord Carnarvon at the door of Tutankhamun's tomb, *wonderful things*.

There are preserved spider crabs framed on the wall, seaweeds pressed in a book, a case of what might be flight feathers or razor clam shells, blown eggs couched in cotton wool, obscure specimens in cork-stoppered jars, plant cuttings laid out on blotting paper, things kept under glass bell jars, and I know, I *know* I should find this all depressing, but I don't, I can't – I think it's wonderful, I think it's magnificent.

There are 1,300 'pieces' held at the Harmas, but that hardly tells the whole story – a recent inventory of 82 plant bundles collected by Fabre, for instance, identified *14,000* different specimens. Fabre was known as an entomologist, and of course there are drawers full of beetles and butterflies, but it's obvious that he was interested in *everything*.

There's a term I've come across that describes the frenzy in which a writer can find themselves when they're seized by the urge to cram everything into a book, to grasp and set down the whole lot, all of it, all of *life*, in words, on paper: 'everythingitis'. I think most great naturalists have a form of everythingitis. I think most children have it, too.

The great nature photographer Eric Hosking recalled a childhood of searching around in hedges and on beaches: 'I commandeered anything in the natural history line, and what I found I wanted to keep.' An early phrase of his, spoken at the age of two or three, was: 'Wouldn't it be awfully decent if we could catch some beekles and put them in a mashbox.'

We kept tadpoles, and then froglets, for a few springs. I caught and fed and eventually hatched a privet hawk-moth caterpillar (a huge fat thing, all eyespots and spike). We made a wormery, pinching sand from the building site at the top of the road. There wasn't that much else. But I played at being a naturalist in other ways. I kept scrapbooks and made pamphlets, wrote booklets, drew worksheets copied from, I don't know, Young Ornithologists' Club magazines, that sort of thing; for a while, one summer, I established what I thought was a sort of wildlife laboratory on the back lawn, a ring of bricks with my nature books set out on them, until my dad said I had to move it so he could cut the grass; I was fond of my *kit*, my third-hand binoculars that my mum had painted green, my little set of plastic jars with magnifying lids. Not just natural history but *being a naturalist* – that was what I wanted.

It was Gerald Durrell's fault. Durrell, of course, was one of the great boyhood collectors; for him, a mashbox of beekles wasn't the half of it. Growing up lawlessly in Corfu in the 1930s, the young Durrell assembled a half-wild menagerie of 'magenpies', scorpions, terrapins, tortoises, snakes, gulls and owls – he documented all this in his memoirs, beginning, famously, with *My Family and Other Animals*, which I adored. But *The Amateur Naturalist* is the book that made me the amateurish naturalist I am today. Published in 1982, co-written with his wife Lee, and probably undertaken to fund the inexhaustible Durrell's relentless adventures in conservation and animal collection, it's a book that tells you about wildlife, habitat by habitat, garden, shrublands, seashore, jungle, but much more than that it hammers home this idea of nature as something you *do*. Each chapter begins with a gorgeously assembled array of specimens and kit: an old wasps' nest, a butterfly net, a killing jar, an alga skeleton, a cuttlebone, a gull skull ('picked clean by crows and carrion beetles'), boar droppings, a 'hand lens', a larva tin. My favourite chapter was the final one: *The Naturalist at Home*. 'The naturalist's workroom is a most important place,' Durrell begins, 'and it must be efficiently laid out, since it is here you keep your specimens and conduct experiments that would be difficult or impossible to carry out in the field.' This was all dangerously exciting. There's a section on 'dissecting tools and their uses'; there are neat diagrams of notebooks and index cards, useful tips on display cases, step-by-step guides to studying ferns and fungi and microscopic creatures; he tells you how to keep aquatic insects and butterflies and slow-worms, he tells you, most memorably of all, how to skin and stuff a mouse.

I never did much of this. I never had the patience or the practical skills. I never knew where I was meant to find a length of muslin for a butterfly case, or some mothballs for getting insects out of a bird's nest. If I did try anything it never turned out tidy enough or complete enough. Here, as much as in Corfu, Durrell represented an ideal I could never really hope to meet.

Genevieve might enjoy pressing wildflowers, though ('if you haven't got a professional plant press', you can always use 'an improvised tennis racquet press'). Danny would love a pooter ('very much like a vacuum cleaner for invertebrates').

I don't know if either of them would be into stuffing a dead mouse – no, I do, they would, they would both *love* it – but there's lots to do before we get to that.

They might turn out to be naturalists, but it doesn't matter if they don't. They might fill our home with insects and reptiles and birds and small mammals, but it doesn't matter if they don't. They'll find their own way into it all – they already are, they're already halfway there – and they'll figure out, soon enough, what they want to *do* about it all. Maybe just watch. Maybe just think. Maybe just care. That's fine, too.

I sometimes feel like I came to nature arse-about-face, because I began with books.

Charles Waterton, writing in 1838, warned against the new breed of naturalist who spends 'more time in books than in bogs'. This was me – not, of course, that Waterton had in mind such books as *The Young Birdwatcher* or *Let's Go Birdwatching!* or the *Usborne Spotter's Guide to Birds* (the biggest natural history book of 1838 was Gideon Mantell's two-volume, 795-page *The Wonders of Geology: Or a Familiar Exposition of Geological Phenomena*).

My grandad was the birdwatcher in our family – the only one, as far as I knew. When I was a kid, he was in his late sixties, early seventies. Chronic bronchitis meant that he couldn't get about much then; by the time I was ten he was seldom out of his armchair. He'd talk, a bit, about the things he'd seen, in arcane places – a firecrest at Grange-over-Sands, a marsh harrier at Martin Mere. He might as well have been talking about volcanoes on Mars or the moons of Saturn, but it was something, it was a start. What was more – what was *better* – was that he had books.

The bookcase in my granny and grandad's house never changed much – even after my grandad died, most of the same books stayed sitting there for years, Granny's and Grandad's together, the *Poldarks*, Georgette Heyer, *The Cruel Sea*, Catherine Cookson, Norah Lofts, the *Cadfaels*. And three wildlife books. I'd guess you could find the same three books on any post-war birdwatcher's shelf. There was *The Hamlyn Guide to Birds of Britain and Europe* (1970), a roller and some bee-eaters on the front, the stiff paperback covers crimped by use. There was the AA *Wildlife in Britain* (1976), each chapter dedicated to a different region of the United Kingdom (by and large I had use for only one of these – Yorkshire, obviously – but I also gave some time to the section on Scotland, which seemed to me then, and still seems to me now, a half-real utopia of eagles and wildcats).

And there was the Reader's Digest *Book of British Birds* (1969), Proust's madeleine, Kane's toboggan, for a certain generation of birdwatchers. You might know it: it's the most successful bird book of all time. Like most nature books of the period, it is largely brown; a tawny owl (which can't be far off life-sized) stares black-eyed from the front cover.

There's no single author, just a long list of expert consultants, presided over by the prolific Richard Fitter. Fitter was an unassuming giant of post-war natural history (Mark Cocker has called him 'one of the great perennials of British wildlife'). It's strange to me to think that he died only in 2005, as his books seem so redolent of a black-and-white austerity-era birding culture, heavy binoculars and flasks of tea, pencil entries in leather notebooks, bicycle rides to gravel pits, *Observer* field guides in the pockets of anoraks – but then it's also strange to me, *too* strange, to think that 2005 was nearly twenty years ago. Fitter's first memory was of watching ducks on Tooting Bec Common in south London; as a boy, he was a keen egg collector. He went on to write more than 30 books, mostly on birds and wildflowers. I have three or four on my shelves: *The Ark in Our Midst*, an ahead-of-its time study of the UK's non-native species; *London's Natural History*, his first book – written while he worked for RAF Coastal Command towards the dog-end of the Second World War and published in 1945 – and the biggest seller; and the *Collins Guide to Bird Watching*, a supplement to his ground-breaking *Collins Pocket Guide to British Birds* of 1952 ('ground-breaking' may sound strong – what Fitter did was to rejig the ordering of the species in his book, grouping them by habitat, size and colour, an ad hoc taxonomy that aimed to be helpful, not scientific – but field guides, for naturalists, are formative texts: a good one can shape the study of natural history for a generation – as, indeed, can a bad one).

I picked up a second-hand copy of the Reader's Digest *Book of British Birds* not long ago. The first thing I did when I got it was open it up and press my nose into it. I don't want to be one of those ohh-the-smell-of-old-books people, but in any case I wasn't after the classic old-book smell, the antiquarian bookshop smell of must, dust, mould, mice, but something a little sharper: I was after the smell of the pages, the chemically treated glossy paper of illustrated 20th-century nature books (and other books, I daresay, but who cares about them?). I don't know how the paper was prepared – modern books don't have quite the same whiff; maybe their pages just haven't yet properly matured, been properly aged – but no doubt it involved coating or steeping in some polysyllabic synthetic polymer or some highly refined mineral soup. Whatever it was, it smells, perversely, of the countryside – it's what wildlife smells like, to me.

In any case, I'm glad Richard Fitter was responsible for the *Book of British Birds*, but it was not so much the text of the book that caught my attention as the artwork. No, so much more than just my attention, it caught my imagination, it *transported* me, in as much as a small boy can be transported while eating Kraft cheese sandwiches and drinking squash on his grandparents' lounge carpet. Eight artists contributed drawings to the book but the full-colour portraits – and the owl on the cover – were the work of Raymond Harris-Ching.

In the late 1960s, the Reader's Digest, in partnership with the publisher William Collins, conducted an exhaustive search of British wildlife painters to identify potential lead artists for the ambitious work that was to become the *Book of British Birds*. None seemed really up to the job – an undertaking they expected to take up to six years and involve the production of 230 colour portraits – until Ching turned up. He was a New Zealander, not long since arrived in the United Kingdom. Here, finally, was an artist whose work had the dynamism they'd been looking for, the vibrancy, the *drama* to snag the casual book-browser (not for them the functional field-guide columns of birds in dutiful eyes-right poses, stiff as

royals, static as the dead, each bird like the last in a different costume). What was more, Ching made the mad promise to do the lot in a year.

The effort nearly did for him – he ended the year ill, exhausted and skint, but my goodness, Raymond, look what you made. Without Raymond Ching, the *Book of British Birds* would still be wonderful, but it would not be magical.

Ching's birds are startlingly alive, irrepressibly *there*. They are as four-dimensional as static art can be, the fourth dimension of course being time, and these birds being captured, snapped, as animals in motion, as animals defined no less by their movement in that dimension than by their length, height, wingspan, shape or colour. There's nothing gimmicky about them – you don't feel that the artist is pushing them into a sort of song-and-dance act in an attempt to wring more life out of them (in fact, a lot of them are doing nothing much at all, because of course living birds often do not do very much at all – they, like Ching's paintings, don't stop being alive, don't lose the spring in their muscles, the light in their eyes, just because they're not moving). Ching's raven, body held horizontal, a tangram of glossy black, lours forbiddingly. His woodpigeon has just seen a field of fresh-sown charlock. His gulls yawp and howl. His swallow, on a wire, has its mind on Africa. His great skua will take you all on.

It isn't fun being told about wonderful drawings that you can't see, so I'll stop there. These, anyway, were the books that made me – or at least, the books that got me started, set me going, like a cheap toy with a dodgy spring, an intermittent connection, an underpowered motor that's prone to malfunction. Since those days, I've been – however off-and-on – a bird person, a wildlife geek, a nature dweeb. Thanks, Grandad.

Actually, there's something else – to do with the *Book of British Birds*, and to do with my grandad – that has informed the way I connect with nature, and therefore some of the ways my kids will do the same. I can rave all day long about Raymond Ching's birds, but the book also captivated a different side of me – the side that is still a dweeb, but perhaps a different flavour of dweeb. It was such a *neat* book. There's no room in the lengthy credits to acknowledge an art director, but I'd like to know who was responsible for the tidy beauty of the box-outs, the maps, the charts ('how long they live', 'British predators and their victims'), the diagrams, the captions, the sub-heads ('the bird-strike menace', 'one mate or several?'). All of bird existence is here: birth to death, wingtip to wingtip, the dance of the grebe, the goldcrest's egg, the snipe's drumming, the osprey's claw, and here I am leaping to my feet to applaud the person who *organised* it. Who tidied it up, shaped it to a 6 in. by 10 in. page, drew lines about it, labelled it, made it fit, made *sense* of it. There is grandeur in this view of life. My grandad was a notably neat man: precise, methodical, disciplined. He was a cricket fan, too. You can picture the Venn diagram – that is, you can see how these qualities and these interests might have some points of overlap. When he was younger, back in Huddersfield, he was a gas engineer. My mum still has the notebooks he produced during his apprenticeship. I'm not sure what exactly they're about, but they're works of such extreme precision, and therefore such remarkable beauty, that I wouldn't hesitate to call them art (though I expect he'd have laughed at the idea). The lines, the graphs, the tables, the columns of figures – I don't think it's just draughtsmanship, something like this, I think it's expressive – I see creativity here, where others might see denial. It's ostentatiously tidy, flamboyantly neat, extravagantly satisfying. I've seen birders' notebooks that are the same way (mine never are, mine are a disgrace).

I expect my grandad liked this about the *Book of British Birds*, too. Perhaps more than the paintings. And I suspect this all comes back to something about the way so many of us have learned to think about nature: as something to be comprehended, listed, logged, measured, photographed, set in columns. Collected. Possessed (though we might fight this part).

So, arse-backwards, I came to books first, and was sold the angle that nature basically makes sense, and is more or less straightforward, and can be explained to you by brisk, genial men with ties and side partings. So much since then, the walks in the woods, the long stints watching the reedbeds, the logs I've rolled over, the rockpools I've poked about in, has, I think, been part of an attempt to map *that* on to *this*, to see the correspondences, to figure out how nature, all the leaping life of it, was ever fitted into a book, how the trick was worked – to find, in short, the 'nature' all the books told me about.

Better, I think, to do it the other way. Learn, first, that nature is a mess, a wonderful shambles, that it can always catch you unawares, won't always do what it's supposed to do, keeps racing forward, a rattletrap at full-tilt, a careering clown car of infinite capacity to surprise and delight, sprawling, magnificent, terrible, beautiful – learn that first, then get down to your reading.

In 'Nature and Books', an arch essay of 1887, Richard Jefferies explained at length the futility of writing on nature. No book or monograph, Jefferies says, can tell him the true colour of a dandelion as it exists in reality. In numerical comparison with the facts of nature, he says, there are no books: 'The books are yet to be written.' A little later, W.H. Hudson, often characterised as an heir to the Jefferies tradition in nature writing, wrote that books appeared inadequate when set against the 'minute history' of even a single bird species. 'It is precisely this "minute history",' he argued, 'that gives so great and enduring a fascination to the study of birds in a state of nature. But it cannot be written, on account of the infinity of passages contained in it.'

Hudson invites us to imagine a young boy who has read 'a dozen long histories' of a given species. Then, on going out to watch this bird for himself, he is amazed and indignant to find how much of what he sees was not mentioned in the books he studied: 'It will astonish him to find how much he has not been told.'

'The reflection will follow that there must be a limit to the things that can be recorded,' Hudson concludes; 'that the life-history of a bird cannot be contained in any book, however voluminous it may be; and, finally, that books have quite a different object from the one he had imagined. And in the end he will be more than content that it should be so.'

———

Lockdown was hard.

All of us, I suppose, paid more attention, gave more time to outdoor things, the woods, the river, the fields, the moors, than we would have otherwise, because in so many ways, through those hard months, it was all we had – the kids couldn't see their friends or play in their grandparents' gardens, still less run around soft play centres or museums or cafés or farms or swimming baths or fairgrounds or community centres, so instead they looked for toads under logs, and for birds in the trees, and for woodlice in the front yard, and for geese in the park, and for the peregrine on the chimney. And I looked at them looking, and I thought about what it meant to them, all this, the animals, the landscapes, nature: what it meant to them then and what it might mean for them in the future – I thought about what it has meant to me, what part it has played in making me who I am, for good or bad, and I thought, too, about what it might mean, in the end, to us all: what it means, why it matters, when we take our first steps into the wild – even when the wild's just down the end of the road, just along the river and into the wood. I think it might matter now more than it has ever mattered.

CHAPTER 1

Chaffinch nest

Hirst Wood, September 2020

SNOOPY: This is the strange creature that was in your
nest? This is an egg! How could anyone not
recognise an egg?

WOODSTOCK: I I I I I I I I I I I I I I I I I I I I I I I I I I I I
I I I I I I I I I I I I I I I I I I I I I I I I I I I I I I I I
I I I I I I I I I I I I I I I

SNOOPY: That's the worst excuse I've ever heard!

WOODSTOCK: Sigh!

CHARLES SCHULZ,
Peanuts, 22 July 1972

P C McGarry arrives at Colley's Mill to ask whether Windy
has seen two boys fishing without a licence. 'If I had,'
replies Windy, a little belligerently, 'do you think I'd tell you?
Weren't you a boy once?'

Windy Miller, as well as absolutely not being a grass, is the
archetype of the English countryman, wedded to traditional
ways (and often contrasted favourably with local agribusiness
magnate Jonathan Bell). Windy 'loves everything about the
country' – we see him fishing, photographing a thrush's nest
and deftly gathering up a bee swarm (and drinking home-
made scrumpy until he passes out, but that's another story).
Camberwick Green was made and originally ran in the 1960s but
it was still being repeated when I was little; despite my ear-
nest efforts to not be a Nostalgia Dad ('put down your Peppa
Pig book, pet, I've downloaded a BBC Acorn simulator'), my

kids watch it on DVD. They may, as a result, absorb the same half-formed idea that I did: something about the lawlessness of nature, something to do with rules and society and the countryside and wildness.

I was a rule-following kid (I'm a rule-following grown-up). The countryside I came to know when I was growing up wasn't obviously lawless; rather, it was a place of uncertain rules, muddy boundaries. Can I walk in this field? Whose path is this? I lived in fear of being yelled at by a farmer ('salt pellets,' other kids whispered warningly, 'they shoot at you with *salt pellets*'). I knew my Country Code, of course I did: I guarded against the risk of fire, I took my litter home, I enjoyed the countryside and respected its life and work; but I could never shake the sense that I might, in some unknown but impor-tant way, be *trespassing*. I have always quailed before a Private Property sign.

Toddlers, of course, are naturally lawless. 'Want it.' 'But it isn't yours.' 'Yes, but want it.' And there they go, toddling off with a fistful of next door's daffodils or the cat's biscuits or whatever it might be. Toddlers are barbarians and enemies of civilisation. This goes without saying.

Take only photographs, people say. No, say the kids (say *all* the kids). We want sticks, sticks that look like swords, sticks that we can ride like horses, sticks for waving and running and going raaa. We want dandelion clocks, we want wisps of sheep's wool, we want poo ('NO Dan' 'But want it!' 'NO Dan it's DIRTY'), we want stones, we want shards of bark, we want leaves, every kind of leaf, we want every kind of berry, every kind of pinecone. Our basic manifesto, the kids say, in so many words, is that we want to find things, and we want to keep the things we find.

We might, on our way through the woods, find a flower, a wood anemone or a cowslip. *Want it*! Hang on, I might say. Let's see. We are on another person's land and are bound by the provisions of the 1981 Wildlife and Countryside Act, which states, Danny, Genevieve, that we may not uproot – that is, dig up or otherwise remove from the land – this plant without permission.

'But want it!' Danny might say.

Genevieve might suggest that we take a little bit of it, a leaf or a flower or –

At which point I might clear my throat and produce a checklist: are we on land designated a nature reserve? Are we on land owned by the National Trust or Ministry of Defence? Are we on land designated a Site of Special Scientific Interest? Is this plant [licks finger, turns page] a red-tipped cudweed, is it a marsh earwort, is it a meadow clary, is it a small alison, is it a Norfolk flapwort, is it a stinking goosefoot ('*You're* a stinking goosefoot,' Genevieve might say), is it a cheddar pink, a tufted saxifrage, an oblong woodsia, a Martin's ramping-fumitory? Is it otherwise a Schedule 8 protected species? No? Then, very well, children: you may pick the flower.

Danny has lost interest and has wandered off into some nettles. Genevieve triumphantly claims the anemone, or whatever it is (I'm not very good on flowers). We'll put it in a jam-jar of tap water on the kitchen windowsill and forget all about it.

Or we might, low down in a thick bush, or in a hole in a beech tree, find a nest and a clutch of eggs. What then?

No ifs, no buts. We must, of course, leave them well alone. There was a sort of sanctity to birds' nests in my mind, as a kid – nests, eggs, baby birds, these were remote and forbidden things, a conception that didn't spring from any internal devotion (I wanted to seize them!), but from relentlessly consistent messaging across kid culture: kids' TV, magazines, bird books, the organs of the Young Ornithologists' Club. There were some things about the countryside that I just came to *know*, as if by instinct: you don't swim in the canal, because the weeds will drag you down; if you see a Colorado beetle, you must inform the government; and you don't go anywhere near birds' nests or baby birds, because no good will come of it. I still feel the awful force of these commandments, down in my bones (if I ever do see a Colorado beetle I will certainly inform the government).

So, we leave nests be. So much about this makes sense, and so much about it makes me terribly sad.

We know, living through this post-war age of ecological nosedive, that we will never see the countryside that our great-grandparents saw. We shouldn't expect to, of course: the one constant in nature is change, and as the human world is reshaped and rebuilt over and again we know that the more-than-human world will, in its turn, be remade too. This is not so much about change as it is about loss. We have lost so much.

Let's take a walk along the river, across the canal, to the woods. It's our walk, one we've walked again and again, just the two of us, me and Catherine, and then me and Catherine and Genevieve, and then the three of us and Daniel, too – our walk, to our woods. It's about a mile. Call the start 1921, when my granny was born, and the end, the gate at the foot of the woods, 2023. Before you've walked a hundred yards the Kentish plover has been lost to England, the black-veined white butterfly, the Mazarine blue: all gone. For economy, I'm omitting the species that have no common name, only Latin (that's most of them, mosses, fungi, liverworts, ferns). The pig's ear fungus and steppe puffball are extinct here by the time you've passed the flower meadow and are in sight of the steps to the bridge, that's 1,500 yards or so. The marsh dagger moth, the union rustic, the red-headed chestnut are on the way out as you cross the footbridge over the river, and as you go up the little path under the trees that smells of wild garlic – you're about halfway there – mosses are vanishing from the British landscape, lesser-curled hook moss, flat-leaved bog moss, sickle-leaved fork moss, helmet-moss, many more. The Norfolk damselfly and the dainty damselfly have gone by the time you get to the road. As you go up past the houses and the derelict plant nursery, this is, what, the 1970s, 1980s, the wryneck goes and then the red-backed shrike, the chequered skipper butterfly, carrion moss, matted bryum. The black-backed meadow ant and the great yellow bumblebee vanish from the United Kingdom just about where you cross the lock on the Leeds–Liverpool. Then it's a short stroll across the car park and into the shadow of the trees. Here's the gate, and here's the wood, and to carry on the metaphor, it's dark in there – there are a lot more losses ahead. All of our remaining reptiles, whales and dolphins, 57 per cent of our amphibians, 43 per cent of our freshwater fish, 37 per cent of our land mammals and seals, 35 per cent of our bumblebees and 33 per cent of our butterflies are depleted or at risk. And the thing to remember is, these were only the *species* losses, the UK extinctions – we still need to talk about decline, so you have to imagine, as you walk, a falling silent of farmland birds, turtle doves, skylarks, of commoner birds like starlings and tree sparrows, of flying insects like bees and beetles; and you have to imagine a dying away of hedgerows, grasses and wildflowers (between the mid-1930s and mid-1980s, about 97 per cent of the UK's wildflower meadows were lost).

We've lost so much *abundance*. This is what I find so sad: that there used to be so much – so much that we took it for granted, of course we did, how could we not, take the egg, uproot the wildflower, what's the worst that could happen? – and now, in comparison, there's so little, and we have to walk so carefully, and feel so worried.

Don't touch. Step away. It's the right thing and the wise thing, I know that. I wish we could take these things for granted, that the larks will nest here again next spring, that the bluebells or harebells or ragged robin will flower again, but we can't – we can't now, and we can't ever again.

We used to find nestlings in the garden or on the drive. Too small for me to know what species they were: sparrows, I expect (sparrows used to teem in our gutters and roof spaces back then). Each just a few wrinkles of pink skin and a wide beak and two bulging eyes, closed, pale grey-blue. Usually they were dead, but not always.

I remember one spring my mum took a pair of them in. They must somehow have both fallen from the same nest (heaved out, possibly, by a domineering older sibling). They were tiny and limp and blind and doomed. My mum put them in bowls lined with cotton wool and propped them on the warm top of the gas fire. I remember my dad called one of them Nermal, after the cute kitten in the *Garfield* comic strips.

My mum collected worms from the garden, blitzed them up in her liquidiser, and fed them to the nestlings through a dropper. Maybe they lasted a day or two.

I remember this in part because of the blitzed worms – it was the same liquidiser she'd use to make Yorkshire pudding batter on the Sunday, but nobody cared – but also because the nestlings were among the very few wild animals that I ever got close to as a kid. Close enough, I mean, to feel their warmth, and their weight (you can't really understand birds until you know that they weigh *almost nothing*).

My childhood was full of books by vets' daughters and farm kids called things like *A Badger Under the Bed* or *Five Owls and a Heron* – I was captivated by the idea of wild things about the house, not exactly tame but tame enough, scampering on to the kitchen table to forage at breakfast time, sparring amiably with the household cats, just being *there*, close enough to touch.

There was the time a jackdaw smacked into next door's bay window and stayed around for a while afterwards, chattering in a dazed sort of way. There were the froglets we raised in an old goldfish tank. There were the back-garden hedgehogs I took some pride in gently manhandling (what you do is, you slip your hand under their tummy, and they squeeze up around you like a tight leather mitten). I touched a python once at London Zoo, but that doesn't count. I must have picked up the odd lost blackbird fledgling and moved it out of sight of the cats. Of course, there were beetles and caterpillars and centipedes and butterflies (I used to try to catch butterflies in my cupped hands, until my grandad spotted me doing it out of his window – red admiral, buddleia – and told me to stop being a – what word did he use? Twit? Wally? I know he said 'imagine fancying yourself a naturalist and trying to catch butterflies like *that*'). But I think this was the sum total of the close tactile contact I had with wild things as a kid. For me (and I can't be alone in this) wild things were defined by being, for the most part, not there – by being just out of sight, just out of reach, going too quickly, always being somehow, uncannily, where I was not. I always felt like an outsider, an interloper, which was fair enough, because that's what I was.

It's bad to disturb birds' nests, it's bad to cause harm, the good thing is to tread carefully, to leave well alone, to keep your distance – I understand all that. I don't want to collect birds' eggs; I understand that the urge to do so is a rapacious, selfish and predatory one; I feel nothing but contempt for the grown men who gather up clutches by the dozen, by the hundred, store them under their sad beds, boast on the dark web of their robberies, pore in secret over their eggboxes filled with blown – emptied-out, dry, dead – harrier eggs, eagle eggs, grebe eggs, phalarope eggs, whatever else.

And yet there is a thread from them to me.

Charles Waterton was – as well as an explorer, taxidermist and first-of-his-kind conservationist – a keen tree climber.

He climbed trees to investigate birds' nests. 'There seems to be an erroneous opinion current concerning some birds,' he wrote, 'which are supposed to forsake their eggs if they are handled.' This, he contended, was not entirely true – as long as one moved 'in gentleness and silence', one might 'take the eggs out of the nest, and blow upon them, and put them in your mouth if you choose, or change their original position when you replace them in the nest, notwithstanding which the bird will come back to them'. Waterton was climbing trees – and for all we know putting birds' eggs in his mouth and taking them out again – into his eighties; he argued (as he always did) from experience. *This* is what I've always wanted: not the eggs, but the experience – not the nests, but the know-how. Helen Macdonald has written about the working-class rural communities in which kids routinely skipped school 'to practise forms of natural history that bent or broke the law: they ferreted rabbits, collected eggs, broke into quarries, kept pigeons, reared finches, climbed fences to poach for fish. Today, they can still spot a linnet's nest in a furze bush at 50 paces and possess fieldcraft skills that would put many a birder to shame.' That kind of close knowing, that intimate familiarity with wild creatures in their wild places – I've never had that. I just didn't know where to find it. I'd love my kids to have it, but I'm not sure where they're supposed to find it, either – I don't think it's as easy to come by as it once was.

It wasn't just working-class kids. David Attenborough collected birds' eggs as a kid; Peter Scott, who co-founded the World Wide Fund for Nature, did so prodigiously as a pupil at Oundle in the 1920s, and even, with a couple of schoolmates, wrote a book about it: *Adventures Among Birds*, by 'Three Schoolboys' ('It was such a beautiful egg that I took it, although I had one already,' runs a typical passage).

The ornithologist Edward Wilson had a close connection to Scott, though they never knew one another. Peter Scott was only two years old when in March 1912 Wilson died at the side of Scott's father, Captain Robert Falcon Scott, pinned down by an Antarctic blizzard at 79 degrees south, twelve miles from safety, 10,000 miles from home, out on the Ross Ice Shelf. Wilson and Captain Scott were old friends (Scott's arm, in death, was found slung across Wilson's chest). He had joined Scott's expedition to the Pole primarily as a naturalist – he was a highly skilled wildlife artist as well as a well-regarded biologist (at the time of his death he was heading up a government committee on grouse disease). He too had started his field studies early, as a boy in the Gloucestershire Cotswolds of the late 1800s. It's obvious from his notes that he was a gifted observer, but for him, too, mere *looking* was not enough.

'Took a hedgehog home – it was put in the vinery to eat beetles'; 'Caught a mole by watching the earth move over its run'; 'Wren caught hold of my finger when I put it in her nest'; 'I caught a mouse by the hind legs this morning'; 'Did you hear the poor little hedgehog is dead? I shall try and get the skull of him. I think it will be a very pretty one'; 'When I come home [this to his mother, from prep school] the house will be run over with all sorts of creatures'. This wasn't a boy who wanted to *watch* nature; Wilson wanted to be up to his neck in it, wanted his hands and pockets full of it. His father, an amateur naturalist himself, said that young Edward had the 'passion for collecting which is the true foundation of the naturalist'. Wilson's biographer, George Seaver, argued that, in taking up wildlife painting, the adolescent Wilson was finding a way 'to make these nature studies more intimately his own'.

Wilson was a collector, in the best sense, to the very end. On that long final march from the Pole in the bitter Antarctic spring of 1912 (Ranulph Fiennes has called it 'the greatest march ever made') Wilson and his companions – Scott, 'Birdie' Bowers, Lawrence Oates, Edgar Evans – man-hauled sleds carrying not only their tent and kit but also, remarkably, 30 lbs (nearly 14 kilos) of geological specimens from the Pole. I've known long-distance hikers cut the handles from their toothbrushes to save on weight. *They have stuck to everything,*' wrote an amazed Apsley Cherry-Garrard, one of the recovery party, in his diary. 'It is magnificent that men in such case should go on pulling everything that they have died to gain.'

Notwithstanding the practicalities of his biological work – and the necessities of polar exploration (Wilson was the party's chief 'flenser' (don't ask) of penguins) – Wilson had by this time come to take a dim view of birdsnesting.

'Much as egg-collection means to me as a collection of reminiscences,' he wrote, 'it is a permanent record of a cruelty I have come to hate in myself as well as in others. I am more inclined every year to leave a nest exactly as I found it.'

And yet in July 1911 – a few months before the start of the expedition to the Pole, and in the middle of the sunless South Polar winter – Wilson, Bowers and Cherry-Garrard embarked on what the latter called 'the weirdest bird's-nesting expedition that has ever been or ever will be', and, more starkly, 'the worst journey in the world'.

They were after penguin eggs. Ornithologists suspected at the time that the embryos of the emperor penguin might offer a clue to the bird's evolutionary history, and, thereby, a glimpse at the earliest bird forms, evolved in the wake of the birds' divergence from the reptiles, deep in evolutionary time. The problem was that emperor penguins lay their eggs on the Antarctic ice, in the Antarctic winter, when travel was next to impossible, and – as Wilson found – temperatures fell to $-77\,^{\circ}$F ($-60\,^{\circ}$C).

The Winter Journey from the expedition's semi-permanent camp at Cape Evans to the penguin rookery at Cape Crozier took nineteen days; the trip back took sixteen. In his memoir, Cherry-Garrard more or less threw up his hands at the impossibility of describing the horror of it: '[It] would have to be re-experienced to be appreciated; and anyone would be a fool who went again.' Frozen sweat, frozen breath, everything (clothes, sleeping bags, harnesses) frozen absolutely solid; crevasses and frostbite, blizzards and pressure ridges, a tent ripped away by the wind; utter physical exhaustion, sleeplessness, and the ever-present prospect of a painful death – all in profound darkness, with damp matches and blubber stoves, for five weeks. They returned, Cherry-Garrard, Bowers and Wilson, with three penguin eggs.

I wonder how often, in the course of those godforsaken five weeks, Edward Wilson thought of springtime in Gloucestershire. *May 9, 1891: Kestrel – 2 eggs from an old Crow's nest in an oak. Jay – 4 eggs, from the end of a spruce bough. May 13,*

1891 – Up at 5, got 3 Magpie eggs. I wonder how often he thought of it as they ferried those three eggs, pickled in alcohol, across the ice barrens of the Ross Shelf.

Later, on the polar journey, hauling blindfold in harness after the glare of sun on snow brought on painful snow blindness, he certainly thought of it; he recalled in his diary imagining that the *shush* of his skis through the snow was the sound of his feet brushing through fallen leaves in the woodlands of Crippetts, Cranham, Crickley, Birdlip: 'I could almost see and smell them.'

Scott, in his final letter to his wife, Kathleen, famously urged her to 'make the boy interested in natural history if you can. It is better than games.' Peter, born in September 1909, just nine months before his father's last ship sailed south from Cardiff, later wrote that 'I cannot remember a time when I have not been interested in natural history'. He was to become one of the most important figures in 20th-century British nature conservation. We may have Edward Wilson to thank for that.

Kathleen Scott did as her husband asked. 'She did not thrust the subject down my throat,' Peter recalled, 'but put me most subtly into the way of naturalists and biologists of all kinds, many of them famous men.'

Of course, it's helpful if one happens to be acquainted, as Kathleen was, with naturalists and biologists of all kinds (Peter described his mother's second husband, the war hero and MP Lord Kennet, with warm approval: 'He was quiet and brave and knew about birds. What more could there be?')

David Attenborough tells a similar story. In 1933, when he was seven, he met a young woman called Miss Hopkins, the daughter of a dignitary who was staying with Attenborough's father (Frederick Attenborough was an academic and principal of University College, Leicester). Young David thought Miss Hopkins 'the most beautiful lady I had ever seen', and boldly invited her to see his 'museum' – a collection of snakeskins, eggs, nests, fossils and conkers, arrayed on a conservatory shelf in the family home. Miss Hopkins consented and listened seriously as she was shown around the exhibits (Attenborough's first audience, perhaps). A few days later, a parcel of treasures arrived: a pearly nautilus, a medieval coin, cowrie shells, sherds of Anglo-Saxon pottery, a dried pipefish. Miss Hopkins – later better known as the archaeologist Jacquetta Hawkes, author of the eccentric landscape classic *A Land* – wondered if David might like to add them to his museum.

But the people who open up the wild to us need not be famous men or women. Natural history is an enthusiasm that, I think, is best passed from hand to hand. I was a bookish sort of kid, and I became a bookish sort of naturalist – but there are, as W.H. Hudson said, dimensions in nature that even the best books can't quite get hold of, and elements in nature study (the knowhow, the *craft* of it all) that are very hard to put into words. Besides that, there's the enthusiasm itself, the infectious joy of birdwatching, botany, the moth-trap, the fungus hunt, whatever – you have to be a hell of a writer to get that to shine out of the pages of your book.

The first person I remember who did this for me was an egg collector.

I must have been about eight. Let's call him Bill. Bill was a regular in the clubhouse at our local sports club (he'd been a formidable hockey forward, I gathered, as a younger man). He was the loudest person I had ever met. In fact I've still to meet anyone louder. He was loud and big and he knew about birds. What more could there be?

When he found out I was a birdwatcher he took me for a walk around the clubhouse grounds. It wasn't the most promising place: there were two grass hockey pitches, a tennis court, a bowling green; at the bottom of the hockey pitches the land sloped steeply away to scrubby farmers' fields, lonely semi-rural pathways and the intersecting railways at Horbury Junction. 'Look here,' said Bill, and, pushing aside the resistant greenery of a privet with one hand and hoicking me into the air with the other, he showed me a song thrush nest: four turquoise eggs in a neat mud cup. He said *there's kestrels, and a little owl*. He showed me where the owl liked to perch. I kept an eye out for the kestrels every time I was there (I was a goalie, which gives you a lot of time to watch the skies). That was it.

It wasn't much, I suppose. But it's been 35 years and I haven't forgotten it. When we were done, he popped across the road, into his bungalow, and came back with an egg from his chicken coop. *Tell your mam to fry it for your tea.* I didn't say 'mam'; I wasn't as broad as Bill. But anyway, I did as he said.

Later I learned about his egg collection – the ones, I mean, that didn't come from his chicken coop. I didn't and still don't know when it was collected (it's been illegal to collect wild birds' eggs since the 1950s, and illegal to possess illegally taken eggs since 1981), and I don't know anything about Bill's ethics, his methods of collection, the species he targeted, any of that. I don't know how big the collection was, but I know it was big.

My main memory of him will always be the four thrush eggs. They were the first wild birds' eggs I'd ever seen; I haven't seen many more since. I tread carefully. I keep my distance.

There were others who helped me find ways into nature. There was my grandad and his books, and my mum (who didn't drive, and heroically schlepped around south Wakefield with me and my binoculars anyway); there were always helpful wardens at nature reserves and birdwatchers in bird hides who'd let you have a look through their scope, tell you what to look for, tell you (more often) what you'd just missed, why you should've been there *yesterday*. And then – because I was a child of the TV Age (1936–present) – there were those who did their inspiring from a distance, those who apprenticed us in wildlife at one remove. There was Attenborough, of course – who'd been in the business for 30 years, even then – and there was Johnny Morris and his avuncular zookeeper routine on *Animal Magic*; there was Su Ingle and Michael Jordan on *Wildtrack*, with its theme tune – which revisits my head about once a week, on average – lifted, incongruously, from John Barry's *Midnight Cowboy* soundtrack; most memorably, there was Terry Nutkins, striking with his shoulder-length male-pattern tonsure and missing fingers, presenting alongside his tame sealion on *Animal Magic* and heading up a crew of younger presenters (including a peroxide Chris Packham) on *The Really Wild Show*. If I think about the nature television of my childhood, it's his voice I hear first.

Nutkins, famously, had a mentor of his own. As a kid in London, he'd skipped school to help out in the elephant house at London Zoo; in his early teens, he made a deeper commitment, heading to Sandaig in the West Highlands to work as a resident assistant for Gavin Maxwell, author of the memoir *Ring of Bright Water* and (among much else) an aristocrat, shark-hunter, racing driver, traveller and poet. Nutkins was to help in caring for Teko, one of the otters with whom Maxwell shared his home.

'Childhood stopped the moment I got on the train to go up to Sandaig,' Nutkins later remembered. Maxwell was a troubled man, a heavy drinker and a depressive. It was a hard place to be a kid: 'I had a life unlike any other boy in Britain ... We climbed mountains. We got shipwrecked. We saw wonderful things. And I became a much tougher and more independent man ... If I knew then what I know now about life and how to handle people, I think I could have done a lot of good up there at Sandaig. I could have helped Gavin. But then I couldn't cope with his intelligence, nor all those days of gloom and whisky.'

Nutkins lost his fingers after a horrific attack by Edal, another of the Sandaig otters (the local doctor who tried to repair the damage immediately after the attack said that it was 'impossible ... like sewing mince'; only urgent care at a hospital on the Isle of Skye – an agonising ferry ride away – prevented him losing his arms to gangrene). Still, he stayed on. Maxwell took responsibility for his education: 'He didn't teach me what the local authorities wanted him to teach but what he thought was best for me – which was natural history mainly. Instead of lessons in maths or science I got lessons about wildlife and life in general and the ways of the world.'

It's an interesting lesson in how engaging with wild things and wild places often, perversely, pushes us into closer engagement with each other. We may define nature as non-human (or more than human), but nature study, natural history, ecology, environmentalism – these are human things, human occupations, and so much of what is good in them lies in our capacity to *share* them. That comes with complications, because people are complicated. But I'm attached to the idea that natural history is a cultural tradition, to be handed on, handed down, passed along with care, and not just a dataset to be downloaded and transferred.

I don't know who'll come to fill these roles in Danny and Genevieve's lives. So far, it's mostly just been us, their mum and dad, heaving up stones for them to look under ('A WOODLOUSE!'), pointing at blackbirds, lifting them up to touch the clusters of berries on the rowans. There are birds and daisies and an apple tree in Grandpa's garden. Granny knows all the flowers.

'You know all the trees!' Genevieve sometimes says to me on our walks, if I've just said: 'This is an oak' or 'This one with the bunches of keys is an ash.' Actually, I know about six trees. At some point she'll learn that there are more than six sorts of tree, and the scales will fall from her eyes.

I know quite a lot of people who know a great deal more than I do about wild things. As the kids grow up, I'll try to make sure these people, the orchid experts and butterfly buffs and people who know where to find lizards and people who understand lichen and people who know maybe seven, eight kinds of tree, get shoved in front of them at regular intervals. I want them to learn about wild things; just as importantly, I want them to learn how to be *excited* about wild things.

Small children, it's true, don't necessarily need a lot of help with getting excited. And besides, there's a lot to be said – I'm saying quite a lot of it in this book, I hope – about looking for yourself, and making your own small discoveries. I remember how it felt when we first had a coal tit on our peanut feeder, and when we first had a siskin (my dad called them Scissetts, Scissett being a village near us, which I thought was very funny, at first); I remember my first treecreeper (among the wet tree trunks and crowding rhododendrons at Newmillerdam, Wakefield), first fox, first newt, first red kite – all trivial things, of course, but still with me, and still mine. When I was thirteen, we took a holiday on the north coast of Scotland (I know the year because we heard of the fall of Mikhail Gorbachev on a broken television, and I learned the word 'coup'). On the very first morning, across from our rented house, on a strip of rocks looking out on the Moray Firth, I saw an osprey drifting over-head. I saw some more a few days later, at RSPB Loch Garten, where, famously, there were and still are nesting ospreys, and there were telescopes trained on them, and information boards where you could read about them, and lovely RSPB people on hand to tell you about them. Loch Garten was (and still is) great, amazing, immensely admirable, but I know which osprey I liked better. *My* osprey. I liked – and I'm not proud of this – that I didn't have to share her with anyone. And I liked the surprise of it, the element of chance.

My mum – so often shanghaied into birding – had an aversion to watching ducks or geese or any sort of waterfowl. If I asked her why, she'd say 'it feels like they've been *put* there'. I know what she meant. I feel it myself, sometimes, in spite of my own good sense, whenever I see something large and unabashed in the wild: a browsing pair of deer, maybe, or an urban fox, or a buzzard on a fencepost. Who put *that* there? It's nice to learn, nice to be taught (though I didn't always feel this way as a child: my parents still laugh about my absolute and visible *disgust* at being shown the right way to hold my cricket bat) – but it's nice, too, to take your own chances, and find your own things.

My middle name is Daniel, and my dad's first name is Daniel (though he goes by one of his middle names, David – I think it's a Northern Irish thing). That's why Daniel, our Daniel, little Daniel, is called Daniel. I'm Danny's dad. There's a high bar for Danny's dads.

'My father, without the slightest doubt, was the most marvellous and exciting father any boy ever had,' writes Danny, in Roald Dahl's *Danny, the Champion of the World* (1975). This is a book about many things, but mainly it's about class and the law, and it's about love and fatherhood. It's also about the countryside. I took in its values, when I was seven or eight, like I took in air. Apart from anything else, I'm not sure I'd call my son 'darling' as much as I do if Danny's dad hadn't done it first.

Danny's dad, William, is a widowed dad who runs a rural filling station. He and Danny live in a well-kept caravan behind the station; their life seems ideal ('Most wonderful of all was the feeling that when I went to sleep, my father would still be there, very close to me'), but it turns out that William has a secret. At night – after Danny has fallen asleep – he leaves the caravan and walks to Hazell's Wood. He's a poacher.

He's also an archetypal countryman – here, again, is the countryside, the authentic countryside, as a place of lawlessness. 'The fields, the streams, the woods and all the creatures who lived in these places were a part of his life,' says Danny. 'I believe he could have become a great naturalist if only he had had a good schooling.'

Danny and his dad look for birds' nests as they walk to school: 'My father told me a nest with eggs in it was one of the most beautiful things in the world. I thought so too.' The poacher, though, has his own moral code: Danny is never allowed to touch the eggs. This makes me think of Bill, of course, and the song thrush nest at the clubhouse (especially as Danny goes on to describe a song thrush's nest: 'Lined inside with dry mud as smooth as polished wood, and with five eggs of the purest blue speckled with black dots.') There's also an echo of another literary dad, this one, too, a sure moral centre in a complicated world: Atticus Finch in *To Kill a Mockingbird*, which I read at secondary school. 'Shoot all the blue jays you want, if you can hit 'em,' he tells his son, Jem, 'but remember it's a sin to kill a mockingbird.'

Drama soon ensues as one night, out poaching, Danny's dad falls into a gamekeeper's pitfall and breaks his ankle. When he doesn't return home, Danny borrows a customer's Baby Austin and goes driving to the rescue (on the way, he voices a classic axiom of the Dahl moral universe: 'Most of the really exciting things we do in our lives scare us to death. They wouldn't be exciting if they didn't.')

The baddie in all this is Mr Victor Hazell, the beery, brutish, pink-faced landowner, who owns the woods and the pheasants who live in it. The law is on Mr Hazell's side, though not the local law, as represented by Sergeant Samways, who, it turns out, is also a poacher: just about every decent person in this book is a poacher, because decency, we're told, between the lines, is not about the law, which will always serve the land-owner, but about doing what you think is *right*, and – and this is important – having wonderful fun in the process.

It's not a bad lesson but it's a difficult one to teach.

Dahl ends the book with a personal message 'to Children Who Have Read This Book'. I remember it from when I was a kid – I think I thought, *lucky me*. And I've read it as a grown-up, as a dad, and it scares the hell out of me.

'When you grow up and have children of your own,' it says, 'do please remember something important. A stodgy parent is *no fun at all*! What a child wants – *and deserves* – is a parent who is SPARKY.'

Well, that's all and well and good, Roald. But how is it done?

There's a marvellous passage in the book where Danny looks at the dull inscription above the door of his school – *This school was erected in 1902 to commemorate,* etc. – and imagines how it would be different if his dad was in charge. Each morning, he says, it would be something new: *Did you know that the death's-head moth can squeak?* Or: *Some bees have tongues that they can unroll until they are nearly twice as long as the bee itself.*

Whenever I try this sort of thing – it somehow became a sub-routine, when Genevieve first started pooping on the toilet, for Daddy to sit beside her and produce a selection of

Animal Facts to surprise and delight – my mind goes mostly blank (often I find I can only summon the most wildly, grimly inappropriate nuggets of information: did you know that when a male bee mates with the queen its innards fall out immediately after ejaculation? Can you believe, my darling, that the mother hyena gives birth through its clitoris, which is seven inches long? In cold winters, my marvellous girl, the great tit has been known to smash in the skulls of other songbirds and eat their brains! I keep these to myself).

Danny's dad is a challenge. I'll never live up to Danny's dad. I'll never live up to *my* dad. The prospect of even trying is a frightening one – but then, of course, most of the really exciting things we do in our lives scare us to death. In any case, I promise I'll have fun trying.

Fantastic Mr Fox is a poacher, too, of sorts – it would be more straightforward to say that he's a thief ('"You are far too respectable," said Mr Fox. "There's nothing wrong with being respectable," said the gentle Badger'). He argues that he only steals poultry from the grisly farmers Boggis, Bunce and Bean to feed his hungry family – but we're pretty clear, I think, that even if his family weren't hungry, Mr Fox would be doing it anyway, just as Danny's dad pinches pheasants anyway, just for the hell of it. But he's an adoring father and husband and a loyal friend (the scene where Badger says 'Foxy, I love you,' jumps out at the modern reader as a sweet and unembarrassed portrayal of close male friendship).

Roald Dahl's English countryside is a land of practical wisdom, bold action and independent moralities, internal and sometimes inscrutable moral laws – I think I knew I wouldn't

have made much of myself there (I'd have wound up a func-
tionary in Mr Boggis' chicken factory or a junior beater on
Mr Hazell's hunt) – which is *why*, of course, I wanted to be
Danny, wanted to be friends with Mr Fox (I think I knew even
then I would never be cool enough to *be* Mr Fox).

Egg collecting – once a decent pastime for more or less
every kid who could climb a tree or scramble through a
hedge – is an interesting ethical weathervane: in so many life-
stories, so many memoirs by nature lovers, conservationists,
we see the vane turn, the compass spin. Hang on, they think,
often on the brink of adulthood, sometimes earlier, paused
over a warbler's nest or a clutch of skylark eggs – hang on,
is this right?

Charles Darwin, as a boy in the early 1800s, collected, but
followed his own sort of code (he was, he writes in his memoir,
made humane by the good influence of his sisters). 'I doubt
indeed whether humanity is a natural or innate quality,' he
writes. 'I was very fond of collecting eggs, but I never took
more than a single egg out of a bird's nest, except on one
single occasion, when I took all, not for their value, but from
a sort of bravado.'

Darwin was one of a number of naturalists from around
his time and not long after who, as they grew older, set aside
not only egg collection but more direct forms of predation,
such as shooting (to a country boy of the 1800s, of course,
shooting birds would have been no more dramatic or unusual a
pastime than hunting out their eggs). Charles Waterton, Gilbert
White and Henry David Thoreau all shot freely as young
men, and far less, or not at all, as they grew older (Richard

Jefferies developed the habit of sighting on a rabbit and then lowering his gun).

I was struck by a moment in a recent memoir by the philosopher Andy West where West's uncle – a cheerfully unrepentant jailbird and career criminal – reflects on the 'egging' expeditions of his youth ('I learnt the coastline by egging. It took me on an adventure. All them old East End burglars and bank robbers you see on the wing in prison, all of us started out egging'). Unexpectedly, he says: 'I regret egging.' Why? 'I was killing unborn birds, wasn't I? I wasn't strangling them, but I was stopping them from having a life.'

It's not how I'd put it – my objections, as usual with these things, have less to do with the harm done to the birds (birds live with harm every day) than with what egging does to *us*, how it indulges our acquisitiveness, our possessiveness, at the expense of wider, deeper, more thoughtful connections with the more-than-human, with our encompassing ecologies – but again, you can see this slow coming-about, this gradually maturing morality. It doesn't have much to do with the law.

Last week, we took Genevieve and Daniel to a farm near Skipton, a lovely place with baby chicks you can stroke and cows you can feed and a real tractor you can sit on and large pigs you can scrub with a yard brush. Genevieve was invited to take a fresh-laid egg from the hen-house – after a moment's pause, she did so, smiling quietly to herself, carrying it from the hen's nesting box to the farmer's basket as carefully as she sometimes carries in the glass milk bottles from the back step. It was the first time she'd seen an egg in any kind of nest. She carried it like she knew what it was worth.

On Genevieve's bookshelf at home there's a book that gives a quite different perspective. I've read it to her once or twice because it's fun to read – it's fun to read because it's by Dr Seuss, but the 1970s eco-messaging of *The Lorax* seems a long way off in *Scrambled Eggs Super!* (1953), a breakneck adventure in 1950s US hyper-capitalism in which a slick kid named Peter T. Hooper ransacks the nests of twenty-odd fantastical birds in order to make some 'fine fancy' scrambled eggs. There's zero message here, no comeuppance, no teachable moment – quite the opposite, in fact, as, by throwing together his plunder – stolen from the ruffle-necked sala-ma-goox, the kweet, the long-legger kwong, the single-file zummzian zuk, the Mt Strookoo cuckoo and the three-eyelash tizzy, among others – Peter T. Hooper succeeds in making his scrambled eggs super-dee-dooper-dee-booper. All this, despite the collection of wild birds' eggs having been illegal in the US since the 1940s.

Genevieve doesn't much care for it. She thinks *Oh, the Places You'll Go!* is better (and she's right).

I want my children to follow rules. Mainly, of course, I'd like them to follow *my* rules (will they *ever*, either of them, not run in the kitchen, use indoor voices indoors, wait when I say wait, put away one toy before getting another toy out – I am a benevolent dictator, these are not oppressive rules, and yet, and yet) – but in general, I'd like them to think that rules have some value, that keeping to rules can do some good. On the other hand, I'd like them to hop over the odd fence or two, go look for wildflowers in someone else's field now and again, take the occasional shortcut across private land (I'll tell them, *choose the place with the snottiest 'keep out' sign*). Land ownership in this country is at odds with so much of what I want to teach my kids, so much of what I want them to believe in – equality of class, equality of access, the equitable allocation of resources, common custodianship of wild places ('the countryside belongs to us all, pet, although actually everything you can see from here belongs to the Duke of Devonshire') – and trespass, as well as being good fun, is a legitimate protest, a decent way of sticking up two fingers to accumulated power. You don't have to be 400 Manchester socialists singing 'The Red Flag' as you leg it up the barren flank of Kinder Scout on your way to a set-to with the Duke's gamekeepers (as happened in 1932). You don't have to be an angry crowd of Lancastrians storming the fenced-off moorland slopes of Winter Hill near Bolton (1896). You certainly needn't be Will Self, joining a 2016 trespass at London's City Hall to protest against the privatisation of public space ('It constrains you in how you think about what you can do in a space,' Self said, 'and it constrains your imagination').

I think you can just be two little kids in someone else's wood. Just being there, that's your protest. That's you staking your claim. *This is mine*, you're saying, even if in a technical sense it isn't, because in another, more important sense it bloody well is.

We found the nest in early autumn – long empty, long abandoned, and presumably shaken from its bush or tree by a September storm. It was lying in the leaf litter in Hirst Wood, still mostly intact, a bit saggy, like a sort of coarse, loose-fitting cap. Moss, grass, feathers, wool. I wasn't sure what bird's nest it was, or rather *had been* – now it was just a clump of stuff (and a superstructure for microorganisms); I ruled out wren (not domed) and song thrush (not lined with mud), but I had to consult bird-minded friends on Twitter for a positive ID. Chaffinch, was the consensus. There aren't many chaffinches in our woods but there are enough.

Back in mid-April – Danny would have been, what, ten weeks old – the female chaffinch started work on her nest. It would have been built in the fork of a shrub: best guess, around here, holly; it would have started with a timber frame, twigs woven into a solid cup; she might have gathered gossamer, spider-silk, to bind the twigs more firmly together. Then moss, grass, wool, for insulation and comfort. This all takes around 1,300 trips, to and fro. Just the right length of twig; just the right sort of moss. Being a bird is exhausting. Birds are exhausting.

This, mind you, is all the female's work. The male stands by – *standing guard*, or so he says.

She probably laid just the one clutch, this year, in this nest. Four or five eggs, pale pastel and irregularly blotched, of which two or three probably died, either in the egg – there are a lot of magpies here, a lot of squirrels, a lot of jays, a lot of great spotted woodpeckers – or after hatching, before reaching adulthood. Then that was it. Everyone left. Just this nest, then, dark and derelict in its fork in the holly.

Dead nests always remind me of an Edward Thomas poem from 1914, 'Birds' Nests':

The summer nests uncovered by autumn wind,
Some torn, others dislodged, all dark,
Everyone sees them: low or high in tree,
Or hedge, or single bush, they hang like a mark.

Since there's no need of eyes to see them with
I cannot help a little shame
That I missed most, even at eye's level, till
The leaves blew off and made the seeing no game.

Thomas is pleased, at least, that the nests weren't found and despoiled by 'boys' (he was writing, of course, in the age of egging), though he can't be sure they weren't got at by jays or squirrels. Then he thinks of the deeply hidden nests he has found, into which berries and leaves have fallen, where he has found the leavings of a dormouse's meal, and where, once, 'grass and goose-grass seeds found soil and grew'. I like, apart from anything else, that Thomas uses the same name for goose-grass, the sticky plant *Galium aparine*, as I do (and the kids do, as they grab fistfuls of it and demand I make it stick to their fleeces, leggings, socks), when he might just as well have called it cleavers, bobby buttons, gollenweed, sweethearts, kisses, sticky willy, claggy meggies or robin-run-the-hedge. And of course, I like the imagery of renewal, the sprouting of clean new grass from this sad thing, this dead thing that was once a home.

Is it a sad thing? I've got it in front of me now. It's a little beaten-up but it's still quite lovely. It fits comfortably in my cupped hand.

Maybe some predator did get to it. Maybe the young all perished – though it was an unusually warm, sunny spring that year, so there's grounds for hope – or maybe the parents fell prey to a sparrowhawk or a cat (chaffinch chicks, like most baby songbirds, and indeed like baby humans, are born helpless, and remain helpless for a long while after). Then it's a sad thing. A few years ago, a mink set up home in my local nature reserve. It's a small site; nowhere to run. Mink can wreak havoc in a place like that. The moorhens and the mallards on the pond kept trying – little ping-pong ball eggs would keep appearing in the moorhen nest at the foot of a flag iris – but it was no good. I kept seeing smashed white shells on the bank. I kept counting the mallard chicks, one day four, the next day three, the next day two, the next day one, then all gone. Mink have to live too, of course – mink have baby mink to feed. But still. Then there's the blue tits over the road, with their nest in an airbrick in the old college building, year after year the same nest, and I *think* the same birds, though who knows, maybe one or maybe both died at some point and were replaced, but, by some Ship of Theseus, Trigger's Broom mechanism, they're still *our* blue tits.

One year a magpie sussed out where the nest was, and how to get into it. There's a famous bit in Laurence Sterne's 1759 novel *Tristram Shandy* where a character dies and Sterne just inserts a single black page in the book. I'd do that here.

But maybe a few of the baby chaffinches made it; what the hell, it was a sunny spring, let's say they *all* made it, let's say they all survived and grew and fledged and learned to fly and are out there now, in our woods, smart in their crisp plumage, going *pink* in the trees, staking out their own territories, making their own nests, their own families. The nest's still empty but I can't see it as a sad thing now – like a discarded snake's skin or the little black batsuit of a dragonfly nymph, or for that matter like a broken eggshell, it's a sign of growth. Like the sacks of baby-sized T-shirts and dungarees and vests we've given away to charity shops or to friends with newborns (keeping the odd thing back, the odd much-worn sleepsuit, the odd dinky sock, we're not made of stone), the nest is something that's simply been outgrown. It is, if nothing else, better than the alternative.

For quite a long time we weren't sure whether we'd be able to have children – a long enough time to get, if not quite *used* to the idea, then at least fairly familiar with it; long enough to sit with the idea, turn it over, see if from all sides (and conclude, in the end: *no thanks – no, we don't want it to be like this, thanks*). I remember where I was while I was on the phone to our GP, talking through my results, count, motility, all that (the doctor said, 'The good news is, you've got *some* sperm,' which I see now is a helpful observation but which at the time felt sort of like waking up after a car accident and hearing a surgeon say, *good news, you've got* some *limbs*). I was in Park Square near the centre of Leeds, a smart green rectangle of grass ringed around with black railings, parked cars and lawyers' offices. I was standing on the grass watching a family of song thrushes while we talked. It was late winter – it must have been a breeding pair and their grown-up children, and the little park must have been their winter territory (there must have been a fruiting tree there, a rowan or a cherry, I don't remember). I don't suppose I was thinking very clearly. I remember having a lot of passing thoughts – no, racing, tumbling, scrambling thoughts – about 'family'. I remember thinking how it's strange that in English we don't have separate words for the family above, so to speak, and the family below – our parents and siblings and aunts and uncles on the one hand, I mean, and our own kids, if we have them, and our niblings, if we have those, on the other – we just have that one word, *family*. I remember trying to figure out which of the thrushes were the parents and which the kids, the adolescents. I don't remember where I went afterwards. I think I had a Don Marquis book with me, an Archy and Mehitabel paperback.

I never felt that our house was empty without children in it. It was impracticably big for just me and Catherine and the cat (it's not a *big* house, but it has rooms under the roof and a big cellar, more than we needed, even with an office – which is now Danny's room – and a room for our books and music – now Genevieve's), but I never felt it was empty. Two people can fill a house. *One* person can fill a house. They can leave all the rooms empty and they can still fill it, just with themselves. It isn't about how much, how many.

Mainly I wanted to have children because I wanted to be part of a family. We're back to the family above and the family below. I was always part of a happy family, my mum, dad, my big brother, my grandparents, my aunties and uncles – I've always been lucky. But the thing about families above – *up*families, can we call them? – is that they don't last forever. So we make *down*families, and become upfamilies ourselves. And of course, we don't stick around for all that long either. After a while it's Trigger's Broom again – the same family, just made up of a bunch of different people.

We cut it a bit fine with our nest-building. We were slapping up Genevieve's wallpaper when Catherine was eight months pregnant, deep into the 2018 heatwave. Were we worried about tempting fate? Maybe we were (or maybe we were just, as my mum would put it, slack set-up).

About one in ten birds' eggs fail to hatch. That's in stable populations of widespread birds – among more threatened species, the figure can rise as high as sixty-odd per cent (it's 65 per cent in Californian condors, 61 per cent in kākāpō). So, on average, that's about one egg in every blue tit clutch – it just fails, no predation, no accident, no rainstorm or spring blizzard, it just doesn't happen. On some occasions this is because the egg hasn't been fertilised (generally due to a male fertility problem), in which case it's as inert as a supermarket Class A. In other cases, the embryo has died in the shell – microscopic studies, where scientists use fluorescence microscopy to look for sperm and embryonic cells within an egg, suggest that this is more common than was once thought. Where this happens at a late stage, with the chick on its way to full development, environmental factors are usually in play: the size of the egg, perhaps (smaller eggs offer less nutrition, have less energy packed into the nourishing yolk), or the temperature in the nest. Where it happens early on, it's more likely down to genetic factors – the embryo is, to use a ghastly term, *not valid*. When we had our first miscarriage, in the winter of 2016, the out-of-hours GP we saw said this to us: *usually, in cases like this, it just wasn't going to happen anyway*. Catherine, at this point, was still pregnant, as far as we knew; we'd come in because Catherine was having pains, cramps – something just wasn't right. The first thing he did, this GP, after we'd told him the problem, was open a browser window on his desktop, type something in, and then slowly turn his monitor toward me to show me what came up: the words *spontaneous abortion*.

Thirty-six hours later, in the early morning, we were crawling through an inch of snow along the B6269. Still dark; the snow, I think, still falling. Red tail-lights strung out all the way up to Toller Lane. At (I think) about 3am I'd woken up and flipped open my tablet to check on the outcome of the US presidential election – it looked like Trump had won. The news was still filtering through as we sat in the A&E waiting room. I heard a doctor in blue scrubs ask a nurse about the election, and then smile grimly and say *god help us*.

We lost our baby that day, at thirteen weeks, in a holding bay in an overstretched emergency department at Bradford Royal Infirmary. We didn't look for answers.

Is it facile, really, to set these things side by side? I can't write about them the same way, their losses, our losses. I don't know how it feels for birds, if it feels like anything at all. I think they know what loss is, at least in terms of a thing that was once there, and now isn't – they may know what grief is, some avian analogue of grief, though I doubt it. They certainly know what change is, sudden, abrupt change, and they know what it is to not look for reasons, not to live with 'why', just to live with the loss and carry on, carrying it with you, but still carrying on (I don't think we *get over* the things that happen to us: I think they get inside us, and stay there, like a nail in a tree grown over with bark; they won't always be painful or even uncomfortable, but they'll always be there).

Those blue tits across our street. To feed their chicks, they might have to collect more than a thousand caterpillars every day. As I said: exhausting. Go out into your garden now and try to find ten caterpillars, try to find *one*. By the end of the breeding season, by late summer, the parent blue tits look shot, knackered, threadbare – but still going.

I'm not going to call it courage, though it looks quite like courage. It's strength, but it's *structural* strength, it's something built in, essential to the basic framework. It kicks in a long while before any eggs hatch, before any hungry gapes open wide, before any baby blue-tit cries of *food, food, food* – that strength, that doggedness, it's both fundamental to parent-hood and there, somewhere in the mainframe, long before the bird becomes a parent. And here I can talk the same way about the birds and us. I think I've seen it in every parent I've ever met – it's something slightly apart from what a person is *like*, and has more to do with who, or perhaps what, they *are*. Think of the old line about the thin person inside the fat person, struggling to get out, or maybe 'The Heavy Bear Who Goes With Me' in Delmore Schwartz's poem of that name (the body, 'a stupid clown of the spirit's motive'), or Kingsley Amis' rueful joke about how having a sex drive is like 'being shackled to a maniac' – all these dualities we somehow intuit between the various forms of 'us', our bodies, our minds, our physical and emotional and rational selves.

Ernest Shackleton, the Antarctic explorer, spoke of a peculiar phenomenon he had encountered while trekking for his life across the frozen island of South Georgia in the South Polar autumn of 1915. With him on the trek – 32 miles, over Antarctic mountains, with no provisions, to the remote whaling station of Stromness – were Frank Worsley and Tom Crean (a veteran of Scott's doomed expedition of 1910–12). Yet Shackleton repeatedly had the distinct impression 'that we were four, not three'. His experience inspired a passage in T.S. Eliot's 1922 Modernist epic *The Wasteland* ('Who is the third who walks always beside you?/When I count, there are only you and I together/But when I look ahead up the white road/There is always another one walking beside you'). And it seems that the phenomenon was not unique to Shackleton: many mountaineers, solo sailors, polar explorers and shipwreck survivors have described the experience of feeling somehow, impossibly, accompanied, in moments of extreme isolation, fatigue and danger. Shackleton's phantom was a fourth, but thanks largely to Eliot this has become known as the 'third man factor'.

Perhaps the parental or proto-parental part of you works something like this – the person in you who keeps you upright, when you're shattered, you've had enough, you're frustrated, you're angry, you're sad and spent and oh fuck this, I can't do it any more, I don't want it any more, can we just not, can I just *stop*, the parent somewhere inside you who might not even say anything, might not even think anything, but just carries on (though of course, sometimes the parent breaks down too: everybody falls, sometimes – and here, too, there doesn't have to be a *why*).

I've seen this in my own parents – perhaps because he worked when I was a kid, long hours, on-call, perhaps because I've seen less in him of other stuff, I've seen it most in my dad. And I've seen it in myself. I've been asked, how does it feel, becoming a parent, being there when your baby's born? And I've bit back the word *duty* – because I can't say that, that thudding, joyless, functional word, 'duty', about the most tremendous, the most wonderful thing that's ever happened to me. But it is the word that leaps up when I'm asked the question. On first seeing them, those messy, bloody, squirming, tiny people, first a girl, next a boy, it wasn't, right away, *my god, I love you*, an endorphin rush, bluebirds singing or a rainbow smiling over the weighing scale – it was something more like: this is me now. This is literally me, this is what I am, and this is what I do. It was, and is, fundamental, in every sense. It doesn't devalue it one bit to say that it feels almost *animal* – that is, it feels so basic, so hardwired, it's not hard at all to imagine that other species, other creatures, feel it this way too.

I've often thought, watching animals, watching birds, how odd it is that we came so late to the truth of our origins – that it took us so long to realise that we came from the exact same place they did. Even now, as a species, we struggle to accept it, but I mean just *look*. Just look at them. It's not anthropomorphism, it's not that I look differently at blackbirds and foxes and mice and sparrowhawks and hares and hedgehogs and herons because they remind me of us, it's that I look differently at us because we remind me so much of them.

Look at sex, look at eating, but above all look at parenting. Sometimes there's something holding us up that's stronger than bone and muscle. All these awful words, 'doggedness', 'resilience', 'persistence', oh god, 'duty' – they're all in there, all part of it, but along with those things, or perhaps within those things, there's such immense and fabulous joy. If this isn't love, I haven't the first idea what love is.

CHAPTER 2

Owl pellet

Rombald's Moor, March 2021

TADPOLE [changing into a frog]: Aagh! Make it stop!

HAPPY: We can't.

NORRIE: It's *nature*.

> 'The Tadpole Badge',
> *Hey Duggee*, 6 October 2016

Nature is a fairytale we never grow out of.

This morning I told my daughter that peregrine falcons eat pigeons. 'Why?' We're in the 'why' phase. They just do, I said. The same reason lions eat zebras. 'Why?' 'They just do.'

We've had peregrines on the mill chimney near our house for more or less as long as Genevieve has been alive. They don't breed there yet but they like to perch up there. We can see them from our back doorstep (and they can certainly see us: a peregrine's eye has eight times the acuity of mine – easily good enough to pick out the Eeyores on Genevieve's pyjamas). For a while she called them 'fairy falcons'. At one point there was a little confusion between 'peregrine' and 'pelican'. She has it right now. She also has a small plush peregrine from the RSPB shop, which lets out a tinny facsimile of the peregrine's shriek when you squeeze it.

Peregrines can – and will – kill anything up to about the size of a mallard. Scientists in south-west England examined prey leavings at three urban peregrine sites and identified 91 wild bird species, three wild mammal species and four kinds of escaped cagebird (including someone's canary). They kill swifts and swallows, woodpeckers and cuckoos, kestrels and sparrowhawks, moorhens and coots, oystercatchers and black-headed gulls, rooks and jackdaws, snipe and woodcock. They kill every kind of garden bird (including the smallest of them all, the goldcrest, which weighs less than a cream cracker and can hardly be worth the effort). They kill bats. And they kill immense numbers of pigeons.

Why?

Well, because they just do. They're made that way. They – and we – are the products of a slow-grinding evolutionary mill that, in coldly sorting benefit and cost, has over many millennia turned out every variety of bird, mammal, plant, insect, sponge, fungus and protist that ever lived. But there's no getting away from it: it's a horrible way to make wonderful things.

We must have been taught about evolution at school. I must have picked up the general idea in biology lessons, in among knotting sausages of semi-permeable membrane (osmosis!) and poking about in lung tissue from the local butcher's. When I finally started reading up on the subject in my late teens – thanks to the pop-science boom headed up by evolutionists Richard Dawkins and Stephen Jay Gould – I was furious. I still am furious. First, you can't understand anything about anything if you don't understand evolution. Second, it's *amazing*. Why, I want to demand, was I not told about all this?

I'll tell Genevieve and Danny about it, of course, as best I can, once they're old enough to grasp the basics. At least, I think I will. The problem is, those basics are really grim.

Growing up is about discovery, but we don't only discover wonders – there is both more and less in the woods than bluebells and anemones, than birdsong and butterflies (dead things in the leaf mould, the silence that falls at the shadow of a sparrowhawk). Genevieve has a jumper that says 'no rainbows without rain' on it, or something like that. It's not a bad summary – but my god, these rainbows, the bluebells, orange-tips, red-flashing woodpeckers, soft green rugs of wild garlic, they need a lot of rain.

I can't see myself sitting down, one quiet evening in the study, Danny on my knee, Genevieve at my side, opening a thick binder and gently going over the numbers with them. *Of all the cheetahs born in the Serengeti,* I might say, *only 5 per cent survive to adulthood, my dears. And those new-hatched turtles we have seen on television, scarpering towards the sea – 999 out of a thousand of them will die before they grow up.* Shaking out a creased academic paper, I might softly read: 'The overwhelming majority of the animals of the overwhelming majority of species appear to have significant suffering but little or no happiness in their lives. Citation: Oscar Horta, University of Santiago de Compostela, 2010. Goodnight, sweethearts. Sleep well.'

Our cat died. I'm writing this on Wednesday. The vet put her to sleep on Monday. Her name was Hedy (after Hedy Lamarr, Hollywood star and torpedo radio-guidance pioneer; we have a minor tradition of naming our cats after female scientists – the last one was Dot, after the Nobel chemistry laureate Dorothy Hodgkin). Hedy's kidneys gave out. She was only seven.

The kids have seen snails squished on the pavement, fruit flies drowned in glasses of squash. They have a sort of an idea – an idea they'll never be able to express, and I'll never get to understand – that there's a difference between a live thing and a dead thing (and a thing that isn't either, a tennis ball or a sock or a drum or a pebble).

Hedy died, we said. *She was very poorly and she died.*

Genevieve cried a bit. Fifteen minutes later she asked if we could get a dog.

It's OK to be sad, we said.

I NOT SAD, Daniel said.

It's complicated. Of course it is.

There's a little hole in our home at present. A little vacancy in the hall corner where before there was a skinny tortoise-shell lapping quietly from her water cup. A silent spot under the bed where before, all through the afternoon, there'd be a soft, comical cat-snore. An empty seat on the back fence. A moment in the small hours where before Hedy would have been wet-nosing my hand and going *grnk*, wanting to be let out – now I'll just sleep through.

Today Danny brightly said, 'Hedy poorly – need medicine!' He still hasn't got quite up to speed with the thing. Genevieve will now and then come out with something like 'we haven't got a cat any more', or 'Hedy hasn't got any friends now' (ulp).

We always had cats when I was growing up. In order of arrival: Sammy (into whose savage fiefdom I was born), Sophie, Smokey and Bandit, Thomas. They all died or disappeared, as cats do, and I don't think I ever felt anything like grief for any of them, except the last, Bandit – but I was grown up by then (he must have been pushing twenty), and Bandit, stuck for the best part of two decades with a stupid pun on his brother's name, was tied up with a lot of stuff about childhood and family and the things you leave behind. So, it's hard to know what the children are thinking. Genevieve's first word was 'Hedy'. Something will stay with them, I suppose. Something will stay with all of us.

She was in discomfort, towards the end; perhaps she was in pain, I don't know. She was quiet, right at the last, on the vet's table. We stroked her chin and scratched her between the ears. Then that was that. So what we told Genevieve and Danny was a story about death, about going from *here* to *there* (what *there* is, exactly, well, we haven't quite straightened that all out with them yet). It wasn't a story about suffering.

When I was a kid, I read a story in one of the James Herriot books that has stayed with me since. Herriot's memoirs of life as a country vet in the Yorkshire Dales are often seen – thanks in part to two charming and largely gentle television adaptations – as cosy and comical, comfort reads, bucolic bedtime stories, and I understand why, because they are warm-hearted and often very funny (they are among the few books my dad has read in full, and when I mention them he tends to chuckle to himself and say something like 'I bet tha's never 'ad to take a cricket ball out of a cow's arse afore, vitnery'). But they are also stories from a serious world – they are about hard lives in the hills, livelihoods that can be ruined at a stroke (long before the harrowing outbreak of 2001, I'd learned from Herriot a cold dread of the words *foot and mouth*). And there's more emotional truth than you might expect.

Andrew Vine visits Herriot's surgery because his fox ter-
rier, Digger, is having trouble with his eyes. The vet diagnoses
pigmentary keratitis – a degenerative condition that invariably
leads to blindness.

Vine abruptly buries his face in his hands.

'I can't stand it!' he sobs. 'If Digger goes blind, I'll kill
myself!'

It's a shocking moment – but what stayed with me (I was
thinking of it as I scratched Hedy's chin for the last time) is
what Herriot says in reply: 'When an animal loses his sight he
doesn't realise what's happened to him. It's a terrible thing,
I know, but he doesn't suffer the mental agony of a human
being.'

He's right: it *is* a terrible thing. And I think the rest of it
is true, too. I believe animals can be far more emotionally
sophisticated than we might expect, and far more intelligent,
and at the same time I don't believe that an animal in pain,
in discomfort, in a frightening situation, thinks *wait, what the
fuck?*, like I would; I don't believe they think a great deal about
their pain and discomfort and fear. I hope they don't. I don't
think Hedy did.

In that paper, Oscar Horta's 2010 paper on 'population
dynamics and suffering in the wild' ('Debunking the Idyllic
View of Natural Processes'), the author does some extraordinary calculations about the lives – and deaths – of cod. The
kids do not know much about cod. There are no cod in their
books. There are no cod in Tynemouth Aquarium and we buy
the cheap fish fingers that are made with pollock, not cod. So,
they do not know much about cod, and they certainly do not

know about the suffering of cod – but if and when they *want* to know, Oscar Horta has chapter and verse.

Horta starts with the proposition that each female cod lays 2 million eggs in the spawning season. In a stable cod population – and thinking hard here about what 'stable' means, exactly, what it requires – just two of the eggs that a female cod lays *in her life* will develop into adults. For convenience, Horta narrows his field to just the million or so adult cod that live in the Gulf of Maine. Giving each egg laid there a 0.1 probability of developing into a young fish (more than a sporting chance) and adding in some speculative figures about how much and for how long the young fish suffer before dying, Horta arrives grimly at the conclusion that 'each time these animals reproduce we can expect that 200 billion seconds of suffering is experienced'. This, he adds, swiftly reckoning up, amounts to 6,337.7529 years of suffering.

Genevieve does not need to know about this. Danny does not need to know about this. When we go up to Heaton Woods and peer into the pond, we'll look at the tadpoles nuzzling the pondweed and I won't say, hey guys, approximately two per cent of frog eggs become tadpoles, zero-point-eight two per cent of tadpoles become baby frogs, and zero-point-one per cent of baby frogs become grown-up frogs, and all the rest die.

That is, however, how all this beauty is made.

Wasteful, Darwin called it: *The clumsy, wasteful, blundering low and horridly cruel works of nature!* Darwin, who saw it all first, who figured out the mechanics, totted up the numbers. He was right, of course. The scale of the waste is beyond our comprehension.

The problem (though *problem* doesn't sound strong enough – 6,337.7529 years!) lies in what is called *r*-selection. This is the reproductive strategy favoured by most of the smaller creatures on earth. The general idea is to produce vast multitudes of offspring and hope that two or three of them pull through to adulthood; to strafe the environment with eggs or spawn or larvae, and then leave them to it – let the law of averages do its pitiless work.

This of course is what cod do. It's not what we do (or what whales, or elephants, or primates, or parrots do). We are *K*-strategists. We produce a far smaller number of offspring and plunge lots of resources into keeping them alive.

Genevieve and Daniel are our *K*-strategy.

It's frightening, really, to think about the resources we do commit to these unpredictable investments, these unregulated securities. I don't mean financial resources (though those, of course, are not insignificant – I think of the character in a Martin Amis novel whose no-good dad presents him with an invoice for his upbringing: '…30 pairs of shoes (approx.) … four caravan holidays in Nailsea…'). I mean, the care, the thought, the worry, the love. Even before they're born, *long* before they're born, we go all-in, emotionally and financially, laying bets we're not at all sure we can afford to lose.

And they are lost, those bets, of course, sometimes – *all* the time, really, if you look at it squarely (I'd sooner not look at it squarely, thanks). We've been lucky, Christ we've been lucky, but we had our losses too, and what's so unfair, what's so *stupid* about the *K*-strategy, is that even when you lose you carry on paying – the investments change their nature, I suppose, the emotions are different, guilt, grief, horror (though I think love remains love); perhaps we wonder *why* rather than *whether* or *if*; but we go on pouring ourselves into them, the lost ones, the absences, the gaps in our lives, giving more to them than we ever thought we had.

It must happen pretty often that a spawning cod disperses her 2 million eggs into the cold waters of the north Atlantic and not one of them comes to anything. They just float out into the pelagic twilight, a cloud, a drift, not growing up, not really *becoming*, just silently submitting to the sea, fish food, free protein, ocean-floor ooze. And the cod, a seasoned *r*-strategist, prolific and oblivious, doesn't have to give a damn. To us, it's profoundly alien (most large vertebrates, in fact, would look askance). But it's how most creatures do this. It is, in a broad but important sense, how the world works. It's how the living universe is made.

The thing about it is, it's an incredibly difficult story to tell.

Not long ago, I was talking to a friend of mine, George, who produces nature documentaries. He asked me what sort of nature programmes I'd want to make, given the talent and a free hand. Ones that reflected the reality, I said, and went on for a bit in much the same vein as above: I don't want to see the little turtle who makes it because most turtles *don't* make it; I don't want to see stories built around survivor bias; I don't want happy endings; I want to be shown how nature really works. So, George said, you'd make a programme that showed unborn and newborn animals dying horribly, one after the other, in an endless parade. Uh, yes, I said. He said he didn't think it was a goer.

The kids' bookcases are full of loving and present animal parents: Fantastic Mr Fox, Fox and Vixen in the Farthing Wood books, Miffy's parents, A.A. Milne's Kanga, Kenneth Grahame's gorblimey dad Otter, a surprising number of owl mums (*Owl Babies*, *A Bit Lost*, *The Owl Who Was Afraid of the Dark*), Fly the sheepdog in *The Sheep-Pig*, the tolerant mummy wallaby in *There's an Ouch in My Pouch!*, Sam McBratney's Big Nutbrown Hare, Beatrix Potter's Mrs Rabbit, and let's throw in Baloo the bear and Bagheera the panther in *The Jungle Book* as pioneering same-sex adoptive parents. Even the Gruffalo seems to be a good dad. It's fair to say that *K*-selection is over-represented in children's literature.

A venial sin of omission, I suppose. Parents spend half their lives in the moral grey area of the white lie. Besides, it isn't as though *K*-selection doesn't have its own heartbreaks: look what happens to Bambi's mum.

I took a break, just now, to hang out some laundry in the backyard. I went out and put down the laundry crate and reached for the peg bag and then had to stop what I was doing for a minute. There was no little cat sitting on the warm lid of the wheely bin, or trotting across the road to be nuzzled, or hiding in the wisteria, or making dissatisfied Marge Simpson noises on the back step. Now this is unfair. This seems really stupid: a misfire of the *K*-strategy, an evolutionary category error, a Darwinian ricochet, an unintended consequence. Hedy wasn't even the same species as me. I wiped my eyes, said *fuuck* to myself, got on with putting the kids' T-shirts on the line.

'Man, they say, is a reasoning animal,' wrote the Spanish philosopher Miguel de Unamuno. 'And yet what differentiates us from other animals is perhaps feeling rather than reason. I have seen a cat reason more often than laugh or weep. Perhaps it laughs or weeps within itself – but then perhaps within itself a crab solves equations of the second degree.'

I finished the laundry and sat on the step and listened to the swift family doing the rounds through the neighbourhood middle-air, four of them abreast. Eee eee eee, they said, on and on. Eee eee eee!

I doubt, myself, that we are the only animals who feel – whatever that means, exactly – but I'm glad that we are animals who feel.

I read them 'There Was an Old Lady Who Swallowed a Fly' yesterday. At the end – *she's dead, of course* – Genevieve looked at me seriously and asked, 'Why do you die if you eat animals?'

'The indifference of children towards meat,' wrote Jean-Jacques Rousseau in 1762, 'is one proof that the taste for meat is unnatural.'

Rousseau, though, was writing around 195 years before the invention of the chicken nugget, 175 years before the spam fritter, 205 years before the Big Mac, 210 years before the Findus crispy pancake, and 217 years before the Peperami ('the ultimate, healthy snacking choice for children'™). Had he lived in the golden age of convenience meat, junk pork, joke beef, in which I grew up, Rousseau might have thought differently. I'm not sure when I first ate meat, but I don't remember feeling any indifference toward it. I was very much a pro-meat child (I was, like many young mammals, an indiscriminate omnivore). At primary school there was no better day than spam fritter-for-dinner day.

In spring last year, the Twitter user 'Average Dad' posted a line from his kid that set me on edge for months. In fact, I'm *still* on edge. 'Dad isn't it weird that the word chicken can mean an animal or a type of food?' his kid had said. *My kid*, remarked Average Dad, *on the verge of making a horrific realisation*.

I don't think Genevieve's there yet. But then 'there' is a complicated place.

Did I know, when I was her age, that spam came – by a circuitous route – from pigs? Until very late in life I believed that *this little piggy went to market* described a consumer piglet that had gone to do some shopping. I absorbed the picture books of Richard Scarry, but don't think I registered the fact that, among the bustling, good-spirited anthropomorphic animals of Busytown, there is a butcher who also happens to be a pig, a smiling, smartly aproned pig, standing beside a platter of bacon, bologna sausage, chops and frankfurters, holding a big knife.

I'm sure I knew, however abstractly, that farm animals got eaten. But there were certain logical steps I just didn't take. Perhaps I wasn't ready to take them. It's quite possible I never will be.

Most of us live our lives some distance from the realities of meat production – that is, of animal slaughter (*meat* is a euphemism: once it was a word that could be used for any kind of food, but now we say it so we don't have to say *animals*). Funnily enough, it was in pre-Revolutionary France – Jean-Jacques Rousseau's France – that European society took its first meaningful steps toward severing the associations between the animal in the farmyard and the meat on the plate.

Eighteenth-century French cooks made it their business to render their ingredients – that is, animal parts – unrecognisable. Cooking became an exercise in *legerdemain* – one minute, a spring lamb, the next, *poof*, a plate of delicious ragout; eating meat became an exercise in denial, or cognitive dissonance. It may be that French fashions for complicated sauces were driven by the relatively poor quality of French meat. The English certainly regarded the whole haute cuisine charade as yet another instance of Gallic treachery. Here, we still roasted and boiled our meat more or less whole, as our medieval fore-bears – or their servants – had done; our food still very often had a face. The scholar Rachel Cairnes argues that by eating animals in something a little closer to their natural state 'the Englishman is a "Brute", a part of the nature he devours'. (Rousseau himself saw it another way: 'The English,' he wrote, 'are noted for their cruelty,' and it was their love of meat that had made them that way. 'I am aware,' he added, snippily, 'that the English make a boast of their humanity and of the kindly disposition of their race … but in vain do they proclaim this fact; no one else says it of them.')

We've lost it now, anyway, we lost it long ago, that intuitive grasp of how animals are turned into food – if we want (and we often do) we can pass through a meat-eating life without ever devouring anything that looks even remotely like a thing that once walked and breathed and farted and fucked and went moo or baa or cluck or oink and ate and drank and had babies and bled and died. Even the words we use don't give the game away: *beef* not *cow*, *pork* not *pig*, *mutton* not *sheep* (this is largely due to a socio-linguistic quirk resulting from the Norman Conquest of England – Old English words for the living animal, Norman French words for the dead).

This is OK, I think, up to point. But I think we may have passed the point up to which it's OK.

When the kids next ask why peregrines eat pigeons, and why lions eat zebras, and why whales eat squid, and all the rest of it, I might look up from my half-eaten pork pie and go, um. Well. Here's another why: why does Daddy eat ham sandwiches, and roast chicken, and sausages, and the occasional burger, and steak, and yes, OK, even, once in a while, a chip-shop spam fritter?

It's not an answer, of course. It's something better: it's a complicated question.

'As a youth,' says the sensitive Uncle Monty in *Withnail and I*, 'I used to weep in butchers' shops.' I never did. I loved butcher's shops. There was always something to look at. When I was a kid in Wakefield, even the city-centre butchers used to have whole rabbits strung up from their red-and-white awnings. You seldom see that now, but I can still kill a happy half-hour walking up and down the butchers' row in Leeds City Markets (even if the market's historic Tripe Shop was forced into closure a few years ago). And yes I *know* they're dead animals and parts of dead animals: that was part of the whole thing, the urgh-yuk-look-at-this appeal, the pigs' ears and trotters and (on a good day) heads, the plucked but unbutchered boiler chickens, the kidneys, livers, lights, hearts. I can see that it's grotesque, and in ways that go beyond the superficial.

There have been some bold attempts, in recent years, to help (or *make*) children acknowledge and come to terms with where nuggets and ham sandwiches come from, what hot dogs and beef burgers *mean*. In the past this would hardly have been necessary, and not only because a greater proportion of British children grew up, back then, around farms and farm animals and the bloody realities of the butchering business. Before the rise of municipal abattoirs late in the 19th century, even city kids, perhaps *especially* city kids, could get a look at the sharp end of the trade, if they wanted, at local urban slaughterhouses, privately owned, but not properly 'private' in any other sense: 'Children hang about the doors,' wrote the author of an 1895 exposé, 'and peer through cracks in the fence, with the usual juvenile delight in sensational developments, but to their own gradual demoralisation.'

The feeling now is that to know is better than to not know – better to let the kids in on the awful secret of what meat is made of than to let them grow up disconnected from the source of so much of what they eat, what helps keep them alive. This is why hundreds of kids on school trips visit the Leverandørselskabet Danish Crown co-operative pig slaughterhouse in the coastal town of Horsens, Denmark, every year – to see what is done, how it is done and, importantly, how it *should* be done, if it must be done at all (Danish Crown prides itself on positive sustainability and animal welfare practices, although, of course, animal welfare stops being an issue pretty abruptly at a certain point). Closer to home – just along the A657 from our house, in fact – is the primary school in Farsley, Leeds, where in 2018 a group of Gloucester Old Spot pigs were adopted by the school farm, with the express intention of having them slaughtered for meat nine months later (the message boards lit up, of course: 'I'm fully behind the pig idea', 'Will the children be watching the slaughter too?', 'They will still be innocent babies when they are murdered and made into sausages', 'There is little point in pretending meat magically finds its way onto Tesco shelves', and so on and so on). An online petition failed; the pigs – never named – duly met their end in the summer of 2019. This was a full decade on from a similar furore over a lamb at a Kent primary school: 'Marcus the six-month-old lamb has now been culled,' Reuters reported, 'after the school's council voted 13–1 to have him killed.'

You can go a little further and explore all the US websites with titles like 'What Children Can Learn From Butchering' (the answers include 'Eating Begins With Death' and 'The Death Begets Gratitude') – it seems a bit weird, of course, until you step slowly backward and look thoughtfully at this whole business, this whole premise of meat and meat-eating, and realise there isn't really a point at which it *stops* seeming a bit weird.

I'm unsure about the idea, widely parroted, that if you eat meat you ought to be prepared to slaughter an animal yourself, and anything else is hypocrisy – by the same lights, I don't necessarily believe you should have to work in a sewer before you're allowed to poo in a toilet. But I do think it's good for kids to be told (and *shown*) the truth. I don't necessarily want to be the one to tell them, or show them, but I think it's good for them to be told. Who knows which way it'll send them, after all.

It seems quite likely to me that in the future, and not even the far-off future, our meat-eating – just the fact of it, let alone the *details* of it, the industrialisation, the corporatisation, the cruelty – will be regarded with horror, like cockfighting or fox-tossing (google it) or torturing bears so that they dance for us. I've had those moments, usually when doing something to a chicken carcass, wresting off a leg or chewing the last meat off a wing or fishing the oysters out with my fingertips or ripping off the crisped skin, as a treat – I've had those *are we the baddies?* moments.

Many people, of course, came to terms with that question, and the answer *yes*, some time ago. In mid-19th century England, Lewis Gompertz, a co-founder of the Society for the Prevention of Cruelty to Animals, was writing cooking notes for vegans ('some persons consider vegetables generally unwholesome, but are wrong ... Pease is one of the best of vegetables ... Celery is also rich and nutritive'); Gompertz also railed against the 'barbarous infantine sports which are encouraged by parents and tutors' – that is, hunting and the like – and quoted a doctor: 'It is absolutely necessary to prevent children from killing little animals, or even to let them see them killed.' In an address to the Pope on behalf of the Animals' Friend Society, Gompertz wrote: 'The impressions made during infancy [stamp] their influence on the character for the remainder of life. We are convinced that the child who commences by pillaging birds' nests is in danger of becoming a robber – that young persons who amuse themselves by killing flies or birds for their pleasure, are, owing to this, the more capable of assassinating their own kind when a temptation may arise.'

Children and idiots, Gompertz notes, contra Rousseau, are generally cruel.

Henry Salt carried on the fight against animal cruelty, against butchery and meat-eating, into the later 19th century and beyond. Salt was influential and well-connected in the progressive and cultural circles of his day – Algernon Swinburne, Rudyard Kipling, Leo Tolstoy, Peter Kropotkin, Keir Hardie and Annie Besant were among his friends and correspondents – and his *Animals' Rights: Considered in Relation to Social Progress* (1892) caused quite a commotion.

The combative and eloquent G.K. Chesterton thought he had Salt with this argument: 'The man who breaks a cat's back breaks a cat's back. The man who breaks a man's back breaks an implied treaty. The tyrant to animals is a tyrant. The tyrant to men is a traitor.'

Salt's early biographer Stephen Winsten put it well: 'G.K.C. thought he had hit upon the basic error upon which Mr Salt went wrong, but it was only a quibble, for tyrant and traitor were equally obnoxious to Salt.'

We are, I'm afraid, the baddies. Meat-eaters like me have become highly skilled at justifying our actions in this respect (see below); I think one of the reasons why we so readily lean into *red in tooth and claw* ideation, why we're so comfortable sitting back and saying 'ah, 'tis nature's way' as another lion rips apart another zebra on the television, is that it makes it a little easier for us to continue as we always have – to carry on being the baddies. I think, at bottom, that when we eat meat we do a bad thing. I understand it – of course I understand it, I *do* it. But it's a bad thing, and our reasons for doing it, for continuing to do it, aren't good enough. The reasons I'll give when Danny and Genevieve ask me *why*: they won't do.

And they're getting worse.

Eating meat props up industries that generate frightening quantities of greenhouse gases. To produce 100 grams of beef protein – the two beef patties in a Big Mac total about 90 grams – turns out, on average, about 25 *kilo*grams of CO_2 equivalent (that is, greenhouse gases, perhaps methane or nitrous oxide as well as carbon dioxide, equal in global-warming effect to 25 kilograms of CO_2). Depending on how the beef is produced, the figure can shoot as high as 105 kilograms. Lamb is bad, too: about 20 kilograms of CO_2 equivalent. In fact, nothing, really, is what you could call *good*: for pork it's about 6.5 kilograms, for chicken around 4.3 kilograms, for farmed fish about 3.5 kilograms. For tofu, soy-derived protein, it's about 1.6 kilograms. Of course, the world is more complicated than this suggests, more full of variables, caveats, loopholes, but in the end it does really come down to the numbers, and the numbers are pretty compelling. One of the most basic and straightforward things you can do to reduce your household's contribution to global warming – to call it by its full name, rapid, catastrophic, irreversible global warming – is stop eating meat. Turn down your boiler, take the train, and stop eating meat.

One of Genevieve's friends at nursery is, she says, a vegetorarian. I should be a vegetorarian, too. I'm not, and I'm not the only one – I'm not the only one who should know better (*does* know better) and carries on anyway.

For me there are – discounting greed, idleness and procrastination (maybe I'll quit next week, or next month; maybe I really will quit when this book comes out, because I'll just be too embarrassed not to) – two main reasons why I still eat meat. One is that I have a lack of basic faith, a sort of profound stupidity, when it comes to the maths of small differences. I know the pennies add up to pounds; I know that all large events result from accumulations of small events, small decisions, small actions undertaken by small people. But I don't, on some important level, *get* it – I have to fight to get completely on board with the idea that my doing *x* or *y*, whatever decent but trivial thing I ought to be doing, *counts*, in any meaningful way. If I do Meat-Free Monday or Freegan Friday or whatever, will tomorrow look any different to today or the day before? I guess the problem is that it won't – it won't look different, but it'll *be* different. I'm not great at understanding that. And I'm not the only one.

There's also the problem of context – the problem that, while I'm spending ten minutes rinsing out a Marmite jar, the US is greenlighting a 629-million-barrel oil drilling operation in Alaska.

Maybe, too, I hate the idea of butcher's shops and family livestock farms going to the wall. Maybe I feel like I'm doing enough – hey, I drive a small and quite shit car and I don't drive it much, I walk to local shops, we never fly, we keep a keen eye on the thermostat, let me have my sausages.

We do what we can, I suppose. That's easier for me to say to myself – *you just do what you can, Rich, you're fine, you're grand* – than it'll be to say to my children. Because it's not good enough, is it? They'll see that right away. And I won't know what else to say.

———

This pellet, then.

I found it at the foot of a stone post on the moors north-west of our house. This is where the coal-seamed mudstone of Airedale humps up to form a ridge of hill country that goes by a variety of names – Baildon Moor, Bingley Moor, Low Moor, High Moor, Rombalds Moor, Ilkley Moor, each shading murkily into the next – until the hump declines steeply into Ilkley and the prosperous valley of the Wharfe. We were on our side, the south side, the Aire side (the cheap side).

Owl pellets are not the only pellets. Any bird that swallows its prey whole – and that includes kingfishers, gulls and crows as well as birds of prey – will at some point bring the less palatable parts back up. Gulls and crows criss-cross these moors all the time; I've seen buzzards, merlins, red kites, kestrels up here, too. What makes owl pellets special is that they are more tightly compacted, and more likely, therefore, to persist for a little while, and to be found on the floor – by, for example, a weary dad taking his kids for a yomp up to Spy Hill on a Monday morning.

Genevieve and Danny were up ahead. They're either lagging miles behind ('I'm just … fixing … these … gravel') or up ahead. I have a hundred photographs of them stepping out in front of me along some winding path or other, in sunhats or raincoats, wellies or shorts, hills and trees and meadows to left and right, ahead of them summits and sky (I'm a bad photographer; I fill all my photos with sky). I saw the stone post, streaked with bird shit, and thought I'd have a poke about in the stiff moor grass at the bottom. You never know. And there it was: a stocky, tapered pellet of grey hair (the same colour as the inside of a hoover bag) with a small jawbone embossed on one side, like a logo.

We were shown these at primary school – I remember someone coming in, bringing owls (I barely remember the owls) and pellets, to be soaked in cups of water and dissected on sheets of pink toilet tissue. I remember, I think, that they didn't smell wholesome. And I remember the fur better than I remember the bones, which seems a bit strange until you think that bones, all being well, are sort of an abstract idea (at least until you take that first trip to A&E with a turned ankle or a popped collarbone and there they are, backlit and livid blue-grey, suddenly – *ow* – painfully real): we're *told* there are bones, and not just bones but whole *skeletons*, inside us, under our skin, and although we take it to be true – just as we take other improbable ideas to be true, ideas like Australia, death, atoms, the 1970s, continental drift, germ theory – we don't really *get* it, not in the same way we get what fur is. Fur is guinea pigs and hamsters and the family cat, we've touched it, stroked it, so perhaps it makes sense that it should have given me pause to find so much of it in something hoicked up from an owl's gullet.

I broke it apart (they aren't very satisfying to break apart, too matted, too hairy) and had a little look through the gubbins within. Tiny bones and rodent hair and insect casings. Owls, even the larger ones, eat more insects than you might think – as with most wild things, their diets tend not to be about what they *want* to eat so much as what they *can* eat.

This, I didn't take home. I let the broken parts fall back into the grass and, dashing the dust from my hands, hurried to catch up with Danny and Genevieve, before one of them pushed the other into a peat pool or they both plummeted into a quarry.

It was stupid, really. There's nothing really gruesome about the dinky vole bones, the stripped shrew components, the remnants of mouse infrastructure that we typically find in a pellet. They've seen worse on the street (a dead blackbird, for instance, baking on the June-hot pavement across from their nursery, a wasp digging head-first into the skin of its neck). They've certainly seen a lot worse on David Attenborough (as small babies, they both watched a *lot* of David Attenborough). They must have seen, at some point, that tragicomic staple of the British nature documentary, the bottom, tail and twitching feet of a mouse being gulped down by a barn owl chick (and I do wonder, by the way, how we would feel about this sort of imagery were owls in the habit of swallowing prey feet-first rather than head-first, the last thing to go not the tail but the mouse's little face, eyes wide and half-comprehending, whiskers quivering, perhaps mouthing 'help' or 'mother!' to camera).

They are both quite scared of owls. Perhaps I had that in mind. A horned owl shrieking unexpectedly at Northumberland Zoo near Morpeth put the wind up Genevieve some time ago and they've never really come back from that. They're OK with *Owl Babies*, with Owl in Winnie-the-Pooh, even with *Hoot Owl, Master of Disguise* by Sean Taylor and Jean Jullien ('I swoop through the bleak blackness, like a wolf in the air') – just not real owls. We've been to a few owl displays for kids and they just won't have it. At one, a barn owl perched on my shoulder. She's been on *Springwatch*, the owner said. Genevieve, unimpressed, had to be taken quietly away, and hugged. I think, at bottom, Genevieve and Danny are both a little afraid that owls want to eat them. They're humans, so they're evolved predators, born to hunt, but at some deep, atavistic level, they also know what it feels like to be the mouse.

Anyway, I didn't take it home. I consider myself fundamentally unsqueamish but it seems that a sort of squeamishness can strike you unexpectedly, from strange angles, when you're a parent.

I think what troubles me most, when I think about explaining things to my kids – when I think about the things I don't *want* to explain to my kids – is not the specific but the general. In terms of the specific, we all have our red lines, our tipping-points, where we think *no, nature, enough – that is too much*. It might turn our thoughts toward theodicy, toward the question of how evil can exist in a world made by a loving god. David Attenborough famously recalled being sent letters asking why his nature documentaries never celebrated or acknowledged the creator of all that splendour, the great whales, the birds of paradise, the flamingos and orchids and butterflies and so on. 'I always have to think too,' Attenborough said, in a 2002 interview with Michael Palin, 'of a little boy sitting on the banks of a river in West Africa who has a worm boring through his eyeball, turning him blind before he's five years old. And I reply and say, "Well, presumably the God you speak about created the worm as well." ... I find that baffling, to credit a merciful God with that action.' Attenborough didn't specify which worm he meant – that there's more than one candidate rather strengthens his argument – but it was probably the parasitic worm *Onchocerca volvulus*, whose larvae are transmitted from person to person by blackfly bites, and which infests riverside populations across sub-Saharan Africa and in parts of South America. A survey in 2017 found that around 1.5 million people living in affected areas had been blinded by onchocerciasis.

There's a dark and angry essay by Mark Twain, unpublished, I think, in his lifetime, and not well known now, that comes to a similar point. Twain imagines a Creator sending the first fly out into the world:

> Depart into the uttermost corners of the earth, and diligently do your appointed work. Persecute the sick child; settle upon its eyes, its face, its hands, and gnaw and pester and sting; worry and fret and madden the worn and tired mother who watches by the child, and who humbly prays for mercy and relief with the pathetic faith of the deceived and the unteachable. Settle upon the soldier's festering wounds in field and hospital and drive him frantic while he also prays, and betweentimes curses, with none to listen but you, Fly, who get all the petting and all the protection, without even praying for it … Visit all; allow no man peace till he get it in the grave; visit and afflict the hard-worked and unoffending horse, mule, ox, ass, pester the patient cow, and all the kindly animals that labor without fair reward here and perish without hope of it hereafter; spare no creature, wild or tame; but wheresoever you find one, make his life a misery, treat him as the innocent deserve; and so please Me and increase My glory Who made the fly.

I like Twain's indignation here, his outrage. It's not clear what lesson he takes, or means us to take, from the fly and its works.

Does he not believe in a Creator? Or does he believe in Him, and despise Him? Twain's argument is especially forceful, I think, because a fly is only a fly, a workaday nuisance, proverbially insignificant – such a small thing, guilty of such small crimes in the larger scheme of things.

Charles Darwin was troubled – in a philosophical sense – by the ichneumon, and I don't blame him: I've had dealings with ichneumon myself. When I was about eight, my schoolfriend Martin and I used to keep cabbage white caterpillars in old wooden drawers. I say 'keep' – there wasn't much to it, you just threw a lot of cabbage leaves into a drawer, threw in a lot of caterpillars and left them to it, it wasn't like tending a rare orchid or breeding giant pandas. There wasn't very much that could go wrong, except that, one day, we were playing with the caterpillars – I don't remember exactly what that involved, probably putting them on things, putting them on other things, maybe trying to get them to fight each other – when I looked down and oops, a caterpillar had gone *splat* on the patio. I thought I must have knelt on it by accident but there was nothing on my knee. The caterpillar had just sort of burst. Its insides were sort of lumpy and a bright piccalilli yellow. They seemed outsized – how had they ever fitted *in* there?

These were, of course, not the poor thing's insides, but the emerging larvae of an ichneumon wasp.

Dan O'Bannon, the screenwriter for the film *Alien*, knew about ichneumon – not only what it does, how it lives, but its horrific imaginative power. 'Parasitic wasps treat caterpillars in an altogether revolting manner,' he wrote, 'the study of which I recommend to anyone tired of having good dreams.'

In *Alien*, of course, the monstrous xenomorph does exactly what ichneumon does, but to a human host, rather than a caterpillar: it lays its egg inside the host's living body and leaves it there to grow and feed and eventually, gruesomely, burst out. This happens to John Hurt at the dinner table. It happened to my caterpillar on our patio. 'The whole notion of [the alien] was taken off a certain kind of insect that will find a host, lay its eggs, and then in that host it will bury its eggs,' said Ridley Scott, who directed the film, 'and then of course the eggs will grow and consume the host. So that's the logic of it all. Probably what makes a lot of nature go around.'

After that first *splat*, that first emergence, it happened every few days. I don't think I ever saw the wasp. It didn't have to stick around long – that's sort of the point.

Darwin couldn't get past ichneumon, couldn't find a way to resolve the ichneumon's life cycle with any sort of Christian concept of 'God'. 'I cannot persuade myself that a beneficent and omnipotent God,' he wrote to the American naturalist and minister Asa Gray, 'would have designedly created the Ichneumonidae with the express intention of their feeding within the living bodies of caterpillars.' He felt the same way about cats playing with mice (which is, I agree, a horrible thing to see). 'The more I think,' Darwin finished, 'the more bewildered I become.' He would be even more bewildered, were he still with us: since Darwin's day, studies have found ichneumon life cycles to be more complex, more remarkable and more cruel than he can ever have feared. Some species target caterpillars while they are still in the egg, so that they are born parasitised; many deploy 'soldier' larvae to wage messy war on *other* parasitic grubs within the host's body, or even to destroy 'rival' grubs from the same parent.

Some naturalists now want the ichneumon wasps – there are around 25,000 species in the family Ichneumonidae – to be given the collective common name 'Darwin wasps'. This feels unfair.

Anyway, how do you explain this, any of this, to a three-year-old? I'm not bothered by theodicy; we're both stoutly unreligious, Catherine and me ('Daddy, is there a god?' 'No.' 'OK.'). But again, becoming a parent has made me squeamish. I can explain the mechanics of the ichneumon, the purpose and use of the long ovipositor, the paralysis of the prey, the growth of the larva – it's horrible but I can explain it. I can explain the fly, explain the parasitic worm. I don't mind all that. But my children have made me, not unsure, not uncomfortable, but, for want of a better word, *unhappy* with the way the universe is. I don't know of any idea of a god that could ever appeal to me (the well-known ones certainly don't) – but oh gosh, wouldn't it just be *nice* if someone knew what was going on? If there was someone in charge, if there was, on a cosmic scale, an adult in the room?

'You see, my dears,' I say, once again gathering my children to me in the flickering lamplight, 'the thing about the universe is, it's all a massive fucking mess.' This is all the philosophy I have to pass on. Also: 'The funny thing about it is, it's a mess, but it can be magnificent. Mag-nif-i-cent. We'll look it up, shall we? Here: "Very grand in size or splendid in beauty: The princess lived in a magnificent palace." No, not a real princess. That's just the example the dictionary gives. No, sweetheart,

I don't know where the palace is. No, there isn't a picture of it. No, listen: what I'm saying is, this mess all around us, it might not look magnificent sometimes. It certainly won't feel magnificent sometimes. Sometimes, you really have to squint your eyes and hold your nose to see it – but it is, I think. I think it's magnificent. I think it's *amazing*. Except when it isn't. And sometimes, my darlings, it isn't. Sometimes it will be rubbish. But it's all we've got. This is it. We make the best of it. We do what we can.'

I don't know how that will sound to a pre-schooler. And it would be so easy to tell them something else, just for now, just till they're old enough to know better. It's not like there aren't options: they're right there, the easy answers, ready-made, waiting, thousands of years in the making. There are stories and songs, there are picture books (*When God Made the World, All Afloat on Noah's Boat!, Baby's First Bible*), YouTube shows; there are cheerful promises, smiling guarantees, all the things that are missing from the way I look at the world. *Just till they're old enough to know better*, though – when is that, exactly? When exactly do we feel ready to confront our big, cold, empty universe? When exactly does any of this start to make any sort of sense? Start trying early is our best shot, I think. Start now.

I'll still tell them the stories, of course. Noah is a good one. Jonah. Daniel and the lions. They'll eat those up. And there'll be other stories, too, ones that I didn't grow up with, from other places, other traditions: the Prophet Muhammad and the baby birds, Guru Gobind Singh and the bear, stories of monkeys, vultures, crows, turtles from Hindu myth. The thing is, I won't tell them they're true.

CHAPTER 3

Jay feather

Moorhead Lane, July 2020

SULA: Uhh, she's so slimy.

BING: Look at her googly eyes!

SULA: Hee hee!

BING: Let's keep her.

'Frog', *Bing*, 17 June 2014

Our feather collection is pretty small. It would be bigger, much bigger, had I not limited eligibility to one feather per species; had I not done that, we would be knee-deep in grubby woodpigeon primaries, feral pigeon primaries, seagull primaries and down feathers pulled from our living-room cushions (species unknown). So, as well as *one each* of those, we have four feathers.

A great spotted woodpecker feather, dark grey with white polka dots. I've seen lots of great spotted woodpeckers, and left to myself it would never have occurred to me to call them 'spotted'. They have some blotchy white business on their wings, but to me they look more like stripes or streaks than spots. The spots on the feather, though – and it's a secondary flight feather, a wing feather – they're definitely spots. I think that this bird, like most birds, was named by a person who had taken one to pieces.

A tawny owl feather. I just took it out of its envelope and had a look at it. It's always smaller than I think it's going to be. It too is a secondary flight feather. Turned sideways it looks very like the body of a tabby cat: dark brown stripes across a light brown ground, petering out into a creamy off-white underbelly. We found it under an elm on our way to the park. I've put it back in its envelope now. It'll still surprise me with its smallness the next time I look at it. Owls, funnily enough, always do the same thing.

A sparrowhawk feather. We didn't find this one. It was detached from a dead sparrowhawk that had flown into a window at my friend Dave's house near Glasgow. 'Something good from the shitness,' he wrote to me as he wrapped it up to post to us. The feather, not the dead sparrowhawk – that he buried under a hawthorn in his garden, 'in hope of an intact skull'. 'It'll be a nice start to the baby's nature collection,' I told him – Genevieve was then just over a year old. So it was.

A jay feather. The prized secondary, a small, barred slash of bright sky blue. Sometimes you see them used in fishing flies or tucked into the tweed trilby of a certain sort of country gentleman. My wife was in the woods with the kids when she found it. For months we kept it looped in one of the straps of the double buggy.

That's four. There was a gilt-edged goldfinch secondary, once, too – a souvenir from the only bird our cat Dot ever brought in. She wore a bell and wasn't much of a hunter. I suspect she found it dead or poached it from another cat. Anyway, she brought it in, singing the hunting song of her people, *ooohhh-waa, mmmmooooh, moowaahh*, and I took it off her and put it first in a sandwich bag and then in the bin, and I kept one feather for a little while, before I decided it was too depressing and threw it away, too.

Small children like to crush feathers. This isn't a sign of disdain or disregard; in the semiotics of small children, smashing things to pieces can be a sincere compliment. Maurice Sendak, creator of *Where the Wild Things Are*, once sent a card to a young boy who had sent him a fan letter, and the boy's mother wrote back and said: 'Jim loved your card so much he ate it.' To love, to want, to seize, to consume – it's all mixed up, for little kids. It takes some time for them to sort it all out. So, feathers, without close policing, get grabbed and squashed and rucked. The barbs get out of line, the feathers get gappy, like broken combs.

Broken feathers are forlorn things. Like broken machines – and what is a feather, if not a fine-tuned flying machine? – they are sad and stripped of purpose. One of their purposes, anyway. It's remarkable that something so precision-tooled for one job should also be so beautiful, and not just beautiful in the way that well-made machines are often beautiful, not just because of their clean lines and efficient proportions, like a jet aircraft or a sports car or a good stovetop coffee pot, but beautiful for beauty's sake. Neither the beauty nor the efficiency is incidental, neither a by-product of the other; both result from millennia of targeted investment.

In my office I have a stuffed sparrow, a house sparrow. It's labelled, in a crabbed hand, *Passer domesticus*, ♀, but it's clearly not ♀, it's ♂ (black bib, grey cap). Genevieve likes to stroke its back: 'It's so *soft*.' The sparrow was a present, and I'm very fond of it (I'm not going to say *him*) – but is it beautiful? It is, when all's said and done, a dead bird on a stick. It's a bit faded but the feathers seem in pretty good nick, considering.

If we accept that a living sparrow is a beautiful thing (and it is), then we have to accept that this almost-perfect facsimile of a sparrow is at least in the same ballpark as beautiful, is at least an 8.5 or a nine to the living sparrow's ten. Unless there's something fundamental missing – some must-have, some dealbreaker that falls in between 'pretty perfect' and 'almost perfect'. Perhaps it's life. Or perhaps it's flight.

In the Natural History Museum in London, in the birds gallery, there's a box of dead hummingbirds. That is, there's a case of stuffed hummingbirds, around a hundred of them, posed in flight, tiny, glittering, pretty perfect – but actually, definitively dead. They would not have died pleasantly, either. They – or at least some of them (the genealogy of the NHM hummingbirds is like *Jarndyce and Jarndyce*) – originated in the collections of the early 19th-century curator William Bullock. Bullock was a rapacious collector, first of artworks, then of 'curiosities', natural history artefacts and, finally, and above all, hummingbirds. He was besotted by their beauty but indifferent to their lives – collecting in Mexico, he amassed almost 70 living hummingbirds, only to let them all die, one by one, in small, dismal, glass-fronted cages. He exhibited a cabinet of 24 taxidermised specimens in 1805, and one of 70 in 1810; in 1819, he auctioned off a collection of 'upwards of one hundred' – at least some of these birds wound up behind glass at the Natural History Museum.

The hummingbird case says a lot about collection – about that part of the collector's worldview which insists that a bird in the hand is worth two, ten, a hundred in the bush, that to possess a living hummingbird would be wonderful, to possess a dead hummingbird would be good, but to possess no hummingbirds at all would be disastrous, a horror, a failure. The NHM recognises this (as in all museums in this age of self-examination and decolonisation, the exhibits at the NHM are as much *about* museums as they are about natural history). And yet, it's still the case that the hummingbird case is extraordinarily, captivatingly beautiful.

A dead bee can still sting. The severed head of Medusa still had the power to turn men to stone.

In 1851, Bullock displayed 24 cases of stuffed hummingbirds at the Great Exhibition at Hyde Park. Among the visitors was Queen Victoria, who wrote: 'It is impossible to imagine anything as lovely as these little humming birds, their variety, and the extraordinary brilliancy of their colours.' And age – some of the specimens must be 200 years old (though in life few survive beyond four or five) – has not diminished them. My friend Jon Dunn, whose book *The Glitter in the Green* is a swooning love letter to the hummingbird family, writes of visiting the museum as a boy in the early 1980s, of stopping at the hummingbird case, and of being 'impaled ... with colour'.

'That they were long dead, shot many decades ago by forgotten men with fowling-pieces loaded with dust-fine shot, and prepared for awkward display by a dextrous taxidermist who had never seen them in real life and had no idea what poses they might naturally strike ... none of this mattered to me at the time,' he writes. 'Here was something otherworldly, something quite extraordinary.'

Raymond Ching, whose bird paintings in the Reader's Digest *Book of British Birds* were so wildly, vividly alive, attributed his own love of birds to an early encounter with a case of stuffed hummingbirds in a New Zealand museum.

No other bird – except perhaps the swift – is so defined by flight as the hummingbird. It can, of course, hover as if motionless in the air, to feed on nectar; it can also move remarkably swiftly (Anna's hummingbird, *Calypte anna*, can hit speeds of up to 50 mph, 380 body-lengths per second, undergoing acceleration of 10G as it pulls out of its display dive). It can fly in reverse. And of course, the hummingbird's flight has its own song.

Something of this persists even after life, even in faded feathers in a gilded museum case – something persists of the ghost of flight.

I just asked Genevieve, would you rather be very beautiful or be able to fly? Without a moment's hesitation she said *fly*. I wasn't sure she would; she's just reaching the age where – from nowhere, through irresistible socio-cultural osmosis – princesses and mermaids and unicorns are suddenly *very impor-tant,* and being beautiful is all at once a thing that matters.

But it cannot – yet – compare with flying.

I asked her why. She said: 'Because it would be really *fast* and *fun* and *high*.'

I think she's right. It would be, too.

Jays can be wonderful fliers. Like most woodland birds, they're extremely agile; life-and-death dogfights with hunting sparrowhawks are not uncommon on wooded heath and in oakwood clearings (jays, being too big to really hide, specialise in the high-risk physical comedy of the last-minute duck, the hair's-breadth sidestep). Over long distances, though, their flight looks less impressive. Jays were actually the first birds I learned to identify from flight alone, thanks to a description in Dick King-Smith's book *Country Watch* (yet another formative childhood bible): the jay, he says, always looks absolutely *exhausted* in flight. They look heavy, knackered, like a plane coming in on a wing and a prayer, pretty much ready to drop out of the sky any minute – but somehow, they never do.

The jay is a beautiful bird – salmon pink, crisp and flickering black-and-white, and that inch of dashing blue – and therefore a decent example of what we will put up with in return for a flash of beauty. The Victorian naturalist Charles Coward said of the kingfisher that it 'is only tolerable on account of the beauty of its plumage' (he considered it a dirty, stupid, idle and intractable bird). The jay is far more worthy of censure: it's noisy, awkward, mischievous, given to stockpiling the skinned corpses of mice and brutish in its treatment of other birds' nestlings. One pre-war writer accused it of 'walking about on any washed clothing the maids may have left on the line'. But goodness it's lovely to look at.

We have a lot of oak trees around here and therefore a lot of jays. When we walk in the woods in early autumn we generally proceed through a light hail of acorns, as the jays (and the squirrels, and the woodpigeons) clumsily ransack the oak-tops.

They were much thinner on the ground when I was a kid. Looking through the old bird books – I mean the *really* old ones, the ones even older than me – there's a running theme, or rather a recurring character, in the entries for 'jay': the gamekeeper. The jay is 'a shady character from the gamekeeper's point of view'; two or three jays are typically strung up beside the stoats, weasels, hawks and owls of a game-keeper's gallows; the jay is 'pursued by gamekeepers with … stupid and deadly animosity'. It's a little chastening to reflect on how much the bird populations of my youth were shaped not only by human intrusion – housing estates, factories, mining, superstores, motorways, all our clumsy ambition, our ham-fisted 'growth' – but by traditions of outright slaughter. Not only jays but peregrines, woodpigeons, magpies, red kites, buzzards – all seem *impossibly* numerous now, compared with a few decades ago. It's not only that there are far fewer gamekeepers. It's that we no longer *think* like gamekeepers.

We are less inclined, now, most of us, to think about wild things in terms of utility and economics, in terms of *good* and *bad*. We think they're all great in their own way – which they are, up to a point (though how would you feel if someone you cared about said that to you? – *I think you're great, in your own way, up to a point*). It's a way of thinking that doesn't always set us up to deal with the hard-edged realities of conservation – what do we *do* when the nests of declining birds are predated by crows and badgers, how do we even talk about that? – but it's had, I think, a transformative effect on the more-than-human world around us, on the living world that my children will grow up in.

Let's circle back to mermaids and unicorns (actually, a speciality of my daughter's, this, inherited from her mum: the conversation that you thought was over, but that has in fact been turned over, reprocessed, thoughtfully reappraised inside her head, to reappear as an abrupt non-sequitur a few hours down the line). We are, we like to think, scientifically inclined parents; Genevieve, we told ourselves, will have no need for unicorns, however pink, however glittery, not when we have all of nature's wonders with which to engage and inspire her questing little mind. Three years on, plot twist: our house is full of fucking unicorns.

I've done a quick stocktake and she has: three unicorn dresses, one plastic unicorn, one pair of unicorn socks, one pair of unicorn underpants, one pair of unicorn wellingtons, two unicorn T-shirts, three books about unicorns (one of which, *The Unicorn Seeing the Ladybird*, she conceived and dictated herself), one unicorn water bottle, one unicorn raincoat and one unicorn board game. Bear in mind she is three and has zero independent purchasing power. Left to her own devices she would wear unicorns like Barbara Cartland wore pink.

We can pull this round. We can save this. We can make this more than a transient mania for fantastical beasts in mauve polypropylene. This is not fantasy; this is *cryptozoology*.

Cryptozoology – that is, the study of living things unrecognised by science, your sasquatches, your yetis, your chupacabras and so on – is largely nonsense and pseudo-science; no accredited college anywhere offers a cryptozoology degree (though you can, of course, study for one online, at – to pick just a couple – the IMHS Metaphysics Institute: '12 Years of Excellence to Humanity', for $1,325, or Thomas Francis University: 'Spiritually-Based Courses and Degrees', for $1,040). There is, however, something worthwhile in thinking about what we 'know' now, and what we 'knew' then – in thinking about what we take for granted, and about how nature can still surprise us. And this has to do with beauty, too.

The Indian wild ass was described around 400 years before Christ by the Greek historian Ctesias. Ctesias was one of the first Europeans to write extensively about Persia and India; his *Indica* offers a colourful – that is, largely fabricated – account of the people, customs, geography and natural history of the region. Among the animals he described is the *Mantichora* or manticore, a creature the size of a lion with a man's face, three rows of teeth and a lethal foot-long sting at each end ('These animals are numerous in India, and are killed by the natives who hunt them with elephants'); he also writes of a large tribe of dog-headed men, around 120,000 strong, who live in the mountains and converse in barks. It's possible that Ctesias wasn't a reliable reporter – but actually, the 'wild ass' he describes may have been real enough. These asses were, according to Ctesias, as big as horses, and sometimes bigger. Their bodies were white and on their foreheads they had horns a cubit in length (so about 46 centimetres). It's very likely that the animal Ctesias was talking about – that is, Ctesias' unicorn, and therefore the first unicorn in Western literature – was the Indian rhinoceros. Now here is an animal I would wear on my socks.

The Indian rhino can run at 30-odd mph, which maps pretty well to Ctesias' claim that the *monokerōs*, or unicorn, can – once it gets up to speed – outrun a horse (of course, specially bred racehorses can run faster than that, but 30-odd mph is about par for a wild horse). It's the largest and heaviest of all rhinoceroses, and like all rhinoceroses it is foul-tempered. Its thick skin is arranged in armour-like folds or plates; its outer incisor teeth are razor-sharp, five inches long, and used – often, and with murderous intent – in fighting other rhinoceroses. Once it reaches adulthood (not easy: 10 to 20 per cent of foals get picked off by tigers) it is about 2.2 tonnes of short-sighted aggression, and essentially invulnerable. That is, of course, until we come along: populations of the Indian rhinoceros have been ravaged by poachers in pursuit of the keratin lump it carries on its face – the unicorn's horn. It's depressing, if not surprising, to read what Ctesias had to say (about 2,000 years ago, remember) about the hunting of this 'wild ass': *Their flesh, being bitter, is unfit for food, and they are hunted merely for the sake of their horns.*

The main point of all this is that the Indian rhinoceros, the real and original unicorn, does not conform to conventional beauty standards. Neither, for that matter, does the dugong, *Dugong dugon*, a big-boned 600-lb sirenian that browses the sea-grass meadows of the Indo-Pacific oceans and is most likely the jumping-off point for the mermaid myth. Dugongs are mammals – they *were* land mammals, but at some point in their evolutionary history they returned to the sea. We think this happened around 60 million years ago, around 120 million years after the appearance of the first mammals, and therefore *long* before the emergence of anything that really looked like a human – but the basic mammal template was in place, and the dugong has essentially the same setup of bones and muscle, and in the right light, seen from a distance, obscured by sea fret and spume, a dugong bobbing among the waves *could* look something like a human (just not necessarily a very alluring human – which is perhaps why many early mermaids, such as the *ningyo* of medieval Japan, aren't very alluring either, often having the top halves of monkeys, or the bodies of fish and only the heads of humans (or monkeys); in other depictions, they are horned, or have mouths full of sharp teeth). It's not much, but most myths are built on not much.

I asked Genevieve which was more beautiful, a unicorn or a rhinoceros. She laughed – *silly Daddy* – and said: 'A unicorn!', of course. I asked her why and she said because it's got *sparkles*.

Beauty came late to these myths. Beauty wasn't always what they were *about*, unicorns, mermaids, it wasn't always the point of them. It certainly isn't the point of rhinos and dugongs.

It's hard to guess what kinds of myths might grow up about us once our civilisation has subsided into ruins and weeds – and then, some time after that, what small children might demand to have on their lunchboxes, socks, hair bobbles and pencil cases. We are living through an age of extinctions, remember; we are living through a time of incalculable and – this is important – irreplaceable loss. Our age, whenever it ends, however it ends, will leave a legacy of great vacancies, gulfs, rifts in the world of wild things. People, if there are still people, will fill these gulfs with stories – that's what people do. Who knows what: perhaps the story of the forest man of Sumatra, a benevolent guardian of the nutmeg and the guava; of the pink river goddess of the Yangtse; of the Szechuan bamboo troll; of the forest mastodon, big as a house, with ears like a tall-ship's sails and a trumpet-call that could shake the birds from the skies, that once, the old folk say, trampled the understorey of the West African jungles.

The stories will make these animals strange and then when people no longer want them to be strange, or need them to be strange, they'll make them *nice;* they'll paint them pink and sprinkle them with glitter. It won't really matter what people do with them, by then; they'll be long gone.

We were down at the community gardens the other day, the raised beds just along from the end of our street where local volunteers plant fruit and vegetables and herbs (for local idlers, like us, to pick and take home). Danny was pointing at the strawberries and shouting 'FRAWBERRIES'. Genevieve was harassing flies. They kept landing on the sun-warmed timber frames of the beds. *Shoo,* she said, leaning close, into the flies' faces. *Go away flies. Go away, shoo.* I asked her, what's wrong with flies? 'I don't like them,' she said. 'They are not very beautiful.'

Well, I am a parent, and therefore nothing if not a hypocrite, and so I read from the approved script: 'Perhaps they are beautiful in their own way, pet – *all* animals are beautiful, darling, if you look at them carefully, with an open mind, consider the plumage of the humble starling, consider the grace of the lowly earthworm…' Anyway, Danny had joined in by now and they were both yelling 'GO AWAY FLIES' at the flies.

In *The Amateur Naturalist*, Gerald Durrell writes of a boyhood encounter with slugs: 'I went for a walk along a mountain road in India with my ayah. There had been heavy rain some time during the day, and the earth smelt rich and moist. At a bend in the road my ayah met two friends, a man and a woman, and I remember that the woman was wearing a brilliant magenta-coloured sari that shone like an orchid against the green undergrowth alongside the road. I soon lost interest in what they were talking about and made my way to a ditch nearby where I discovered, to my delight, two huge khaki-coloured slugs brought out by the rain. They were slowly wending their way along the ditch, leaving glittering trails of slime behind them ... To me, they were not only fascinating but, in their own way, as pretty as my ayah's friend in her beautiful sari.'

This is how I am supposed to think, and it's how I would like my children to think. It is not how I think, or at least not exactly. I do, as it happens, find slugs quite beautiful: a long, clean, glossy black slug stretched across a footpath on a damp day, say, eye stalks at full length, chalky sunlight picking out the ridges of its skin. But living things aren't just for looking at (Durrell, a formidably hands-on sort of naturalist, knew this). Take one of those long slugs in your hand – go on, I dare you. I don't even mind the slime (which is, by the way, not incidental, not an accident or by-product of basic slugness, but the slug's evolved defensive superpower, its equivalent to a snail's shell, a mechanism for gumming up predatory beaks and teeth). The slime is just slime – I'm not going to eat the thing, after all. What I can't stand is the slug's *strength*, the ghastly reflexive muscularity of slug bodies in the hand.

In early courgette season – once the children are in bed, and the neighbourhood slumbers – I have been known to run them through with a sharp trowel. It's like chopping aubergine.

So I'm not a slug fan, not, as a rule, unequivocally pro-slug. I also don't like it when craneflies get all up in my grill, as they invariably do (if it's not what craneflies are *for*, it's certainly what craneflies think they're for). I can handle spiders of all (UK) sizes, but I don't want one racing up the inside of my sleeve. I'm fine with maggots – I remember, when I was a kid, running a hand through the hundreds of writhing maggots in my friend's angling bait-box, and I remember how soft and cool they were, like flour – and I'm on good terms with the multitudinous worms in our compost, but fat, rope-like worms, worms of over six inches in length, nope, no thank you, no worms for me, thanks, ta.

Of course I know, as a matter of general principle, that all living things *are* fascinating, and beautiful up to a point, in certain restricted senses and contexts. I treasure a photograph I took this year of a horse leech, beetle-black and eight inches long, pulsing glossily around the foot of a marsh orchid at a nature reserve near Earsdon, North Tyneside; however, on my office wall, here above my monitor, there's a framed print of an Audubon painting of a blue heron, and for all my self-conscious open-mindedness there's a reason why it's a painting of a blue heron and not a pulsating horse leech. As someone who is *into nature*, I feel it's incumbent on me to make valorous knightly defences of ugly – no, of *not conventionally attractive* – animals. Nature is wonderful, these pullulating maggots are nature, therefore these pullulating maggots are wonderful. There's something – for me – uncomfortably religious about it all.

All things bright and beautiful, the hymn goes. I don't believe in any Lord God, but I'll make sure Danny and Genevieve know that whatever made the bright and beautiful things – evolution, dancing to music of chance – also made the foul and stinking things, the ugly, the weird. Consider the aye-aye: razor-sharp teeth, raggedy hair, huge ears, beady eyes and a long, twig-like forefinger with which it drums and probes for insects. It's a type of nocturnal lemur, native to Madagascar. In a 2009 nature documentary, the writer Stephen Fry characterised the aye-aye as looking 'as if someone has tried to turn a bat into a cat'. It is, he said, 'certainly bizarre, for some even a little revolting. And I say, long may it continue being so.'

The aesthetic philosopher Emily Brady used Fry's personal remarks about the aye-aye as a jumping-off point in her marvellous 2011 paper 'The Ugly Truth'. Brady's own position is that 'ugliness in nature is real' – that it needn't have anything to do with damage or deformity (a healthy wolf fish, 'bulgy eyes, widely spaced teeth, outsize mouth and dull grey colour', is still an ugly wolf fish) and that 'ugly nature ... remains ugly, even if our response is mixed, involving dislike but also curiosity, wonder or fascination'.

This made me think of the work of the biologist and broadcaster Simon Watt. Watt embraces the ugly in nature; he leverages it, turning ugliness into something like a selling point – through the Ugly Animal Preservation Society, through a public vote in 2013 to elect the world's ugliest animal as the society's mascot, Watt pushes the very necessary line that ugly species – blobfish, scrotum frogs, pig-nosed turtles, aye-ayes – must be embraced, if we're to do anything serious about protecting the natural world (the blobfish won the vote, by the way). 'Uglier animals are neglected,' Watt says in the society's mission statement. 'I set up the Ugly Animal Preservation Society as a tongue-in-cheek way of trying to redress the fact that the cute and cuddly species like the panda dominate natural history books and TV shows. By being blinded by beautiful animals, we not only miss out on the joys of hearing about some fantastically weird and wonderful creatures, but we might actually be harming our planet.'

Watt points out that, while invertebrates account for about 79 per cent of animal life, they are only covered in 11 per cent of scientific papers on conservation. 'Ugly animals are less likely to be researched,' he says, 'never mind protected. This taxonomic bias constrains the capacity to identify conservation risk and to implement effective responses.'

That's the serious thinking behind a daft, funny campaign that Watt and his team have taken to schools and kids' science shows, using comedy to make the point that *ewww* is okay, as is *haha* or *urgh, gross* – as long as we don't stop there, as long as we don't let aesthetics shape our understanding of what's important in nature and what works in terms of slowing down our planet's ecological decline.

Brady's paper shows us that there are lots of ways to think about ugliness, beyond those reflexive *urghs*, beyond *go away flies, shoo*. We have found lots of ways to persuade ourselves that ugly things are actually beautiful, to use our heavy human brains to – at last! – think ugliness out of existence.

'The natural environment,' writes the philosopher Allen Carlson, 'insofar as it is untouched by man, has mainly positive aesthetic qualities; it is, for example, graceful, delicate, intense, unified, and orderly, rather than bland, dull, insipid, incoherent, and chaotic.' This is quite straightforwardly untrue: it's quite easy to see that toads are not graceful, rhinoceroses are not delicate, woodlice are not intense (flies are not very beautiful). But we can look at it in other ways.

Another philosopher, Holmes Rolston, argues that *apparent* ugliness becomes beautiful when we consider ecological context – that a decomposing elk carcass, say, ceases to be repellent when we consider its role in a healthy ecosystem, when we think of happy maggots and food chains and population stability and so on. 'The ugly parts do not subtract from but rather enrich the whole,' Rolston says. 'The ugliness is contained, overcome, and integrates into positive, complex beauty.'

Samuel Alexander expresses a similar sentiment, from a more aesthetically minded angle. Alexander likens ugliness to discord in music or horror in a dramatic tragedy – it is not to be considered on its own terms, but as a component of something greater: 'An ingredient in aesthetic beauty.' 'Such ugliness,' he wrote in 1968, 'is a difficult beauty.'

It's an echo of that lovely schoolroom pabulum, *life's rich pageant*. 'Knocked a tooth out? Never mind, dear, laugh it off, laugh it off; it's all part of life's rich pageant,' says the games mistress in a celebrated 1937 monologue by Arthur Marshall. An older usage, by a newspaper journalist in the early 1930s, well articulates a child's-eye view of our complicated world: 'Left to himself, the average child relishes a street fight or an accident. To him they are not shocking, but a thrilling part of life's pageant – inexplicable but existent.' *Inexplicable but existent.* I like this very much. I like it because it says nothing about beauty. Replace 'street fight' with 'decomposing elk' and we have a neat rebuttal to Dr Rolston. We can *relish* the ugly. We don't need to bend over backwards to find things beautiful; things are allowed to not be beautiful.

We're in our pale-blue Datsun Bluebird and the sunshine through the trees is dappling the road ahead. My dad's driving. My mum's in the front and my big brother's next to me. Either my mum or my dad will be smoking a cigarette (Mum's were B&H, Dad's were Regal). I'll have my window open. My granny and grandad, with my auntie in the back, will be following behind – most likely some miles behind – in Grandad's bottle-green Ford Popular. We're headed east. We're going on holiday! We always go to the east coast and we usually go to some place with a curious name that, to me anyway, will always just mean 'holiday' and nothing else, a place that outside two August weeks in the early-to-mid-1980s will simply not exist, a Gransmoor, a Chapel St Leonards, a Scalby Mills, a Primrose Valley, each a summer-holiday Brigadoon, conjured up along with herring-gull yawps and bladderwrack smells when we arrive, vanishing into the sand and thrift when we go home, unquenchably sunlit, always The Holiday House, just ours, no, just *mine*, for ever.

Some parts of our childhoods are, if not exactly universal, then at least held in common with thousands of other childhoods: almost all of us, once we're a little older, can find things to remember together, fragments of school or family life, TV, books, music, films, juvenile adventures, adolescent fears or feelings or misapprehensions, the common stuff of kidhood, shared in – generally without realising *anyone* else was in on this – with kids across the world and through-out time.

And then there's *your* stuff and *my* stuff, and we're allowed to hold this close, and keep it to ourselves, because it really is just ours. Lots of families had Datsun Bluebirds. Lots of mums smoked B&H. There are lots of big brothers. There have been lots of summer holidays. But no one else was on that back seat going to those places with my family and the sunshine through the trees dappling the road ahead and my grandad following behind and our Flanders and Swann tape playing on the car stereo.

I suppose I have to dab my eyes at this point and explain to younger readers (those born after the last interglacial) that Flanders and Swann were an English stage duo celebrated for their comic songs: the brilliant and convivial Michael Flanders wrote the words, and Donald Swann, bespectacled, owlish, complex, gifted, played the piano. My grandad had *The Bestiary of Flanders and Swann* – can you believe it, I never got that pun until just now – on vinyl. I remember the sleeve showed a giraffe eating Flanders' hat. We had it recorded on tape (my kids have no idea what a tape is, of course, and why should they, but I've just checked and am pleased to report that *The Bestiary of Flanders and Swann* is available on Spotify). I'm not sure how it became our holiday standard, our must-have A64 soundtrack, because family traditions, like fossils, form in opaque and unpredictable ways, but it did.

The *Bestiary* was an album of animal songs – 'The Armadillo' (a romantic number, and still my personal favourite), 'The Elephant', 'The Whale', 'The Portuguese Man-of-War' ('I do not care to share the seas / With jellyfishes such as these') and suchlike.

The first song on the album – 'The Warthog (The Hog Beneath the Skin)' – is a song about ugliness.

I remember, when I was a teenager, and tentatively discovering how to be a wanker, I said to my dad that the *Bestiary* songs were 'sort of surreal', and he said no, son, they're silly, they're just silly. He was partly right. They are silly, but not *just* silly – they're also satirical, often quite sharply so (consider, for example, the chameleon who 'colours himself and his opinions by the company he's in', and the ostrich who buries his head in the sand and gets blown up by a nuclear bomb – not the subtlest, that one).

Flanders obviously knew a bit about animals – from the *Bestiary* you learn, among other things, that it is the male seahorse who carries and gives birth to the young, and that a three-toed sloth is also called a bradypus – and, treading carefully, hearing in my head my dad saying *they're silly, just silly*, I wonder how much, in writing these songs, he was looking inward, at himself, as much as at the snobberies and inconsistencies of the post-war English.

Flanders was 21 and serving in the Royal Navy when, in 1943, he contracted poliomyelitis. He remained in hospital until 1946, sustained by an iron lung that had been specially constructed to accommodate his burly 6 ft 4 in. frame. Later in life, he reflected phlegmatically that 'I do not look upon my years in hospital as the least valuable in my life. Being pruned back in my early twenties and given time to think and study has its advantages.' So, when you listen to the words of

'The Sloth' – in which the idling bradypus reflects on what he might have done if he were not a sloth ('I could paint a Mona Lisa / I could be another Caesar') – it's hard not to think that they were informed by his long years of convalescence.

Similarly, however silly it all is, it seems striking that a song about prejudice and the superficiality of conventional beauty standards was written and performed by one of the very few theatrical celebrities who used a wheelchair.

'The Warthog' tells the touching story of a young female warthog who is shunned at the post-hibernation ball, despite her efforts at beautification. A happy ending arrives in the form of a haughty gentleman warthog who at last asks the heroine to dance – because, despite her fripperies, he can see that she is 'the sweetest little, neatest little, dearest and completest little warthog ... underneath'. I suppose it sails a little close to 'a face only another warthog could love', but still, I found it quite lovely as a kid. Michael Flanders never gave any outward impression of insecurity or fragility (a *New York Times* reviewer reported from Broadway that Flanders 'skates about the stage as though his wheelchair were a swan boat in the process of making figure eights along the road to Valhalla'), and like many people with disabilities considered his impairment irrelevant to his work. Still, as a quietly effective campaigner against discrimination, he surely understood the power and importance of the message. It comes up again in his spoken lyric 'The Duck-Billed Platypus' – we have given the duck-billed platypus a silly name, Flanders says, but the platypus doesn't care, because he doesn't call *himself* a duck-billed platypus: 'He's a Golden Shining Lovebird, in duck-billed platypese.'

Danny and Genevieve haven't been started on Flanders and Swann yet – they were both sung to sleep with 'The Armadillo', but they were too young at the time to formulate a proper critical response – but they have their own window on to the relative ugliness of the warthog.

'Hello! I'm the warthog, as ugly as sin,' says the warthog, arrestingly, in Julia Donaldson's book *The Ugly Five*, 'with two pairs of tusks and a bristly chin. My tail stands on end and my body is dumpy. My head is too big and my skin is too bumpy'. In the book, the warthog joins up with the wildebeest, the spotted hyena, the marabou stork and the lappet-faced vulture to form the foul-looking but mutually supportive quintet of the title (for what it's worth, I don't think the wildebeest belongs here at all, being really just a sort of tapered cow, and really quite a looker when set alongside the marabou stork, which really is a shit-caked monstrosity from hell). The upshot of *The Ugly Five* is that the animals' adorable offspring come rolling out in the final act to tell the ugly five – now the lovely five – how wonderful they are: 'Hello! We're your babies. You give us our food, And help cheer us up when we're in a bad mood' (the ugly parents are also thanked for 'picking out the nits' – this reminds me of a poem I wrote in a Mothers' Day card when I was in primary school, which specifically highlighted my mum's excellence in 'giving the cats their cat food, and keeping their litter clean' – love you Mum, so proud of you, keep on clearing out the cat piss).

We do, of course, get this from our kids, whatever and whoever we are – a sort of unconditional love, even if it might not be *quite* without judgement (Genevieve likes to tell me that 'your hair is not *really* as beautiful as mine'). Really, it's more an unconditional *joy* – joy at the sight of us, our faces, joy at the sound of our voices, regardless of our aesthetic failings. I recognise this from my own childhood.

And I recognise it, too, in my relationship with nature, with wild things. This sort of joy that has nothing to do with beauty at all.

You don't have to look closely to find this. You don't need to philosophise. It's not about proportion or colour. Yes, a bird's egg is a pleasing, satisfying shape; yes, there are gleaming oil-rainbows of colour in a starling's wing, but I don't really *care* about that – it's miles down the list of things that get my attention. It seldom starts with 'ooh, that's lovely'. It might start with 'whoa, that's gross'. It very often starts with 'what the fuck is that?' It has to do with presence, with the *there*ness of a wild thing. Here is something new (wild things are always new, new every day); here is something alive and complicated, something that does things, thinks things, feels things, sees things. Here is *something*. So, attention must be paid.

It's more than easy, it's instinctive. The kids get it, too. They wanted to come see the horse leech (they nearly trampled down the orchid to see it). They want to see, they want to smell, they want to touch; they want to grab it and take it home. So do I. Sometimes I do; sometimes I find other ways (mostly I write about it – that's what so much writing is, just a grabbing and a taking home). Do you want to see a spider? Do you want to see a rat? Do you want to see a dead jellyfish? YES! YES! YES! Do you want to see a warthog? Do you want to see a lappet-faced vulture? Are you mad? *Bring them to us.*

It's like that journalist said: it's life's pageant. Except it's more than that, it's better than that, because life isn't *meant* to be a spectacle, it isn't meant to bring anyone joy, it's just turned out like that; it's just a huge, joyous, messy accident – or rather, nobody planned it, but it's turned out that the world has evolved to fascinate us, and we have evolved to be fascinated by the world, and after millions of years it all comes down to two small children in wellies and splash suits, looking under a log in the woods, poking at a fungus with a stick and filling their dad's pockets with pebbles and feathers.

CHAPTER 4

Limpetshell

St Mary's Island, August 2020

ROBIN: Cockleshell Bay is only a little town – not like
Roughington, where we used to live. And it's
got the sea, and the beach, and boats and that.

ROSIE: Yes.

NARRATOR: Rosie was beginning to feel a bit happier.

'A Fresh Start', *Cockleshell Bay*, 6 May 1980

Take nothing but pictures, people say. Leave nothing but
footprints.

Genevieve's little feet barely make a mark in the wet beach.
The soft dents left by ball and heel are soon enough smoothed by
the regathering sand. We do take things, though. Of course we
do. Our hands demand it. And the sea won't miss them, I don't
suppose – these limpetshells and winkleshells, this coin-sized
carapace of an ex-crab, this black hand of dried-up saw wrack.

Weren't they richer, rockpools, wasn't the seashore busier,
when I was a kid?

Of course, everything seemed bigger. I remember a
Northumberland holiday and crabs the size of dinner plates,
B-movie crabs, ancient crabs, vast and barnacled, lurching
sidelong across the rocks (my dad, an inveterate nicknamer,
called them Cuthberts, for no good reason, and so now I call
them Cuthberts, for my own reasons). I remember visiting our
family in Northern Ireland and on Portstewart Strand – where
the beach was so broad and empty you could drive a car across

129

the intertidal zone, and choughs nested on the cliffs – walking on wet sand cobbled with clamshells wider than my hand, longer than my foot.

We mostly find limpets and winkles, now. It may be that I'm too old to look properly and Genevieve is too young (Danny is certainly too young: he hangs face-down and wide-eyed in the baby carrier as I loom over the pools). Not even the ketchup splots of sleeping sea anemones. Not even a jellyfish washed up by the tide. Still, I'm sure before too long we'll come upon something more memorable: a starfish, maybe, or a whelk or a clingfish – or a little whale beached on Tynemouth Longsands, like in Benji Davies' *The Storm Whale* (one of our favourites), or a 'great big, grey-blue humpback whale', like in Julia Donaldson's *The Snail and the Whale* (also one of our favourites: big fans of stranded cetaceans in our house, it turns out). Still, there are always the seals.

St Mary's Island, just off Whitley Bay, is reached by a tidal causeway and is best known for its lighthouse. We call it Grandpa's Lighthouse (technically, Genevieve calls it *Prampa's Lighthouse*, but she'll get there). Catherine's dad, who lives just up the road, has taken a series of wonderful photographs of it, and one hangs in a frame on our living-room wall; looking at it used to calm Genevieve down when she was little and making a fuss. The first time we came here she said *Prampa built this, wow, it's really big*. It is really big, she's not wrong – 46 metres tall (the same height as the Statue of Liberty, torch to toe) – but it was built in 1898, a little before her grandpa's time. It blinked its paraffin light in the North Sea darkness for nearly a century and was decommissioned in 1984. The main reasons to go inside it now are to count the steps to the top ('…one hundred and thirty-six, one hundred and thirty-*seven*…') and to watch the seals.

They breed further out, and further up, at Coquet and – further still – the Farne Islands. St Mary's is where they come to haul themselves out of the sea and rest. They lounge and bask on the rocky skirts of the island. Here and there you see a round seal head bobbing in the surf, glossy as a wet pebble. Genevieve watches them with her plastic binoculars held the wrong way up.

Suddenly, the seals panic. It's a clumsy, lurching panic, a wave of flight response that rolls awkwardly through the herd of 60-odd animals: whiskery muzzles lift, blubbery bodies twist and flounder towards the breakers. Pretty soon we see why. A wakeboarder comes drifting into view, maybe 50 yards from the shore. Tall and poised and Giacometti-thin in his black wetsuit, he paddles slowly along a line parallel to the horizon, not seeming to see the seals, or, if he does see them, indifferent to them.

There are polite notices propped on the rocks all around the lighthouse, asking visitors – birdwatchers, photographers, intrepid pre-school rockpoolers – not to go any further, not to trespass on the seals' territory. They're so easily spooked, and even up here, where 70-odd miles of largely open coastline stretches from the north edge of the Tyne estuary to Berwick and the Scottish border, there are only so many safe places for them to go.

The wakeboarder has turned for shore, and there are only a few seals left on the rocks, when Genevieve starts getting restless (that is, getting a bit too close to the edge of the viewing terrace and the sheer drop beyond and having to be hoicked back by her anorak hood) and it's time for us to go. 'BYE BYE SEALS,' she says. We're too far away for them to really hear. That's sort of the point.

Finn the Little Seal by the illustrator Sandra Klaassen is the story of a grey seal cub who – with the help of his mum and his friend Sula – overcomes his fear of the 'big sea' ('the big sea is *far* too big for me') and leaves his safe and comfy rockpool behind. It's a story every parent is familiar with, of course. Kids' lives are just one big sea after another – a sequence of brave little jumps into deeper water that, from here anyway, seems like it never ends (*does* it ever end?). Nursery school, pre-school, primary school; new words, new skills, new feelings, new ideas.

Finn the seal finds things in the big sea a bit easier when he realises that a lot of things he'd come to know in his rock-pool – the starfish, the anemones, the hermit crabs – are there on the seafloor, too. It's an analogy for life, because there'll be familiar things in our children's lives that will be there for them no matter how big they get, no matter what school year they're in (me, for instance: I refuse to go anywhere, ever). But there's a literal reading, too. I know that nothing in nature stays the same – that's the *point* of nature – but sometimes, in the woods, on the shore, we can find at least a sense of permanence.

Rockpools are in some ways the most transient of habitats, transformed twice a day by the rolling tide. In other ways, though, they are – as the property shows say – *forever homes*.

Home scars. A good title for a novel. These are the shallow gouges made in marine rocks by the limpets who cling there at low tide, sealed against sunshine, secure beneath their ridged, conical shells – the same shells that now tinkle together in my pocket each time I take off my coat. This is attachment to place in the most literal, fundamental sense. I remember, as a kid, trying to kick them off the rocks with the heel of my shoe (so that we could use them as crab bait) – they are stubborn, impossibly stubborn, and will endure any amount of hammering rather than release their grip. So fierce is their hold that eventually they shape the rock to the footprint of their shells, and their shells, in return, adapt to the contours of the rock. I like that it works both ways. Anything we hold on to tightly is reshaped by our holding on. Everything we love is reshaped by our love.

Acorn barnacles, too, are defined as much by *where* as by *what*. They are 'sessile' or immobile, self-cemented into place (after a short free-swimming larval adolescence, each joins itself permanently to a solid substrate by what, in a barnacle, passes for its forehead). Under a bare foot – or through the sole of a small boy's jelly shoe – they feel, en masse, abrasive, prickly, coarse but not unpleasant, at least until you fall over, as you invariably do, and skin a knee or elbow. There are 120-odd species of sessile barnacle, spread across fifteen families and three suborders. Barnacle study – cirripedology – can be a life's work.

Charles Darwin, for one, devoted years to them. 'I hate a Barnacle as no man ever did before,' he wrote in the autumn of 1852, six years into his painstaking study of the cirripedes and with two years still to go. I imagine that he already had barnacles on his mind when in the mid-1840s he wrote to his old *Beagle* captain Robert FitzRoy that, having established himself and his family at Down House in Kent, 'my life goes on like Clockwork, and I am fixed on the spot where I shall end it'.

So he was. If not entirely sessile – and certainly not cemented by his forehead to a length of driftwood – Darwin was a homebody by both nature and necessity. Chronic digestive problems, never diagnosed, made it difficult for him to spend time away from Down, and so it was at Down, by and large, that he remained, with his wife, Emma, and their children.

It's important that Darwin was a dad – important to me (I don't really have heroes, but if I did, he'd be one of them), and important to any real understanding of who he was. He and Emma had ten children between 1839 and 1856; at one point he joked to a friend that Emma 'has been very neglectful of late & we have not had a child for more than one whole year'. Only seven of the Darwin's children, however, lived to adulthood. Charles' favourite, Annie, died at the age of ten. In their notes and letters – almost intolerable to read – the mourning Darwins recall a kind, sensitive, happy little girl – and Charles remembers, too, the things Annie liked to do: 'She had some turn for drawing, & could copy faces very nicely ... She liked reading, but evinced no particular line of taste. She had ... a strong pleasure in looking out words or names in dictionaries, directories, gazetteers, & in this latter case finding out the places in the Map.' Perhaps she would have become a scientist like her dad (and her brothers George, an astronomer, Frank, a botanist, and Horace, an engineer). Science in any case was second nature to the young Darwins – at Down, experiment and observation were part of the household culture. There's a story about how George, as a little boy, visited a friend's house and learned that the property didn't have a study. 'But where does your father do his barnacles?' he asked.

'The father in Down House *did* barnacles and he *did* bees and he *did* carnivorous plants and he *did* worms,' wrote Rebecca Stott in a *Smithsonian* profile of Down. 'And if the father did them, so did the children. These children were willing and happy assistants to their attentive father, fascinated by his explanations of the natural world. As soon as they were old enough, they were recruited to oversee certain experiments – to observe seeds growing on saucers arranged on windowsills, or to play music to worms, or to follow and map the flight path of the honeybees across the Down House gardens.'

I like to think of Annie playing music to her dad's worms.

Worms are something we have plenty of here, in our little yard (we don't have a lot else, really, besides snails, and spiders, and the odd roving blue tit). Danny and Genevieve both like being tucked under my arm while I carefully lift off the lid of the compost bin. Worms! Percolating diligently through our coffee grounds and cat litter, our vegetable trimmings and fruit skins. Danny can say it: *worms*, and then: *more worms!*

Perhaps next time we'll bring out Genevieve's wooden harmonica and Daniel's little drum. Give them a tune, or something like one. See how they take to that.

Within a decade of Annie's death, Darwin had attained scientific celebrity (and, in other circles, an unrivalled notoriety).

Had she lived, she would have been eighteen when her father finally published *On the Origin of Species* – she would certainly have read it, and it's curious to wonder what she might have thought about it. What will any of our children think about what we think? Even if we never write our masterpieces, never make the world shake as Darwin did; the ideas we have, the view of the world we've cobbled together and piece-by-piece pass on – what will they mean to our children? What will our children *do* with them?

Edmund Gosse was born a few years after Annie Darwin. His father was Philip Henry Gosse, a distinguished naturalist and writer on marine life, and a member of the fundamentalist Plymouth Brethren sect. When Edmund's mother Emily died from cancer in 1857, Philip raised Edmund alone, by the seaside at Sandhurst in Devon.

Philip's two main concerns were his work – the fifteen-spined stickleback, the rainbow leaf worm, How The Whelk Conquers The Bivalve, and all the rest – and the salvation of his son. In his memoir *Father and Son*, Edmund details the years of harrowing struggle between the pair: Edmund's intellectual and emotional growth (he was to become a celebrated *fin de siècle* man of letters), and Philip's painfully intense efforts to protect the boy's soul from the devil, and the world. Philip was the model for the eccentric, stern and stiflingly loving father Theophilus in Peter Carey's novel *Oscar and Lucinda*; an episode in which Theophilus flies into a fury because his son has taken a mouthful of Christmas pudding ('his father said the pudding was the fruit of Satan') is based on an incident recalled in *Father and Son*.

I can't go rockpooling – especially not if Danny and Genevieve are with me – without thinking of Philip Gosse.

Edmund remembers the rockpools of the north Devon shore as 'our mirrors, in which, reflected in the dark hyaline and framed by the sleek and shining fronds of oar-weed there used to appear the shapes of a middle-aged man and a funny little boy, equally eager, and, I almost find the presumption to say, equally well prepared for business'.

The funny little boy was certainly no slouch: in 1859, at the age of nine, toiling together with his father on the shore, he discovered a new genus of sea anemone, the walled corklets. In his book *Actinologia Britannica*, Philip credits 'the keen and well-practised eye of my little son'.

Only on the shore, Edmund writes, did his father's face lose 'the hard look across his brows … that came from the sleepless anxiety of conscience'; here he became 'most easy, most happy, most human'.

The joy Gosse found in those teeming Devonshire waters came to have a painful counterpoint. The wild popularity of his books, beginning in 1853 with *A Naturalist's Rambles on the Devonshire Coast*, helped to make rockpooling a fashionable craze among a Victorian public granted new access to the seaside by the expanding railway network. An 'army of collectors', as Edmund puts it, passed over the rock basins – by the time he came to write his memoir, the beauty he and his father had known was 'long over, and done with'. 'The ring of living beauty drawn about our shores,' he writes, 'was a very thin and fragile one … No one will see again on the shore of England what I saw in my early childhood.'

It was not lost on Philip that his own work had played a part in this despoliation. Edmund writes that the knowledge 'cost him great chagrin'. Thinking again of how the stern and careworn man of God was transformed on the shore, among the corallines and anthea, the seaweeds and anemones, toting chisel and jar, with his boy at his side – most easy, most happy, most human – it feels like an understatement.

Further along England's south-west coast, a generation earlier, another young collector had walked the shifting shore with her father – another seaside kid with a keen eye for curious things. Mary Anning was trained in three things: reading and writing (perfunctorily, by Dorset Congregationalists) and fossil-hunting (by her father, Richard). Richard Anning was a Lyme Regis cabinetmaker who supplemented his small income by selling 'curios' found in the sediment layers of the Jurassic shore: 'devil's fingers' (fossil belemnites, squid-like creatures), 'snake-stones' (fossil ammonites), that sort of thing. He and his wife Molly had ten children: only Mary and her older brother Joseph lived to adulthood (Mary was given the name of an older sister who was killed in a fire in 1798, aged four). Encyclopaedia entries pass off this staggering mortality rate as 'not unusual for the time' – true enough, but think what a world of tragedy and sadness is contained in those words. Richard himself died, from tuberculosis coupled with injuries sustained in a cliff fall, when Mary was eleven. The family, always skint, was now reliant on poor relief – and fossils.

She was twelve when they found the ichthyosaur. It wasn't called that yet. It wasn't called anything yet. Her brother Joseph found the skull, 4 ft long with a big crocodilian grin.

Over the next twelve months, Mary found the rest. A team of workmen was hired to lift the monster from its bed; it was sold, as 'a crocodile in a fossil state', for about £23, and then sold on at auction for around £45 (in between it was owned and exhibited by William Bullock, of hummingbird fame). The surgeon Everard Home called it a proteosaurus; in 1817 it was given the name we know it by today, ichthyosaur, 'fish-lizard'. Nobody thought to name it after the Annings; for all Mary's later fame as a fossil hunter, as the woman who opened up the Jurassic Coast to science and on whose dangerous and poorly paid work a whole new discipline was built, no British scientist thought to name a species after her in her lifetime (only the American naturalist Louis Aggasiz did: two fossil fish were given the name *anningiae* in the 1840s).

It's often said that Mary Anning was the inspiration for the tongue-twisting nursery rhyme 'She Sells Seashells on the Sea Shore', but it isn't true – aside from the various historical inconsistencies, Anning was a specialist: 'I don't know any thing of shells,' she wrote in 1822. What she knew was palae-ontology: *my old passion for bones*. In that same 1822 letter, she lists the things that might be found in the rocks of 'her' coast, the Lyme Regis Blue Lias formation, between Charmouth and Axmouth: 'Ichthyosarus vulgaris, Icthyosarus platydon Tenuirostis, Plesiosarus, dorsal fins resembling the radie of Balistis, dapedium politum, fragments of three other distinct species of fish, Crustacions insects, vegetable impressions, pentacrenite, a variety of Amonites, Natulis, Belemnites, palates of fish Wood, and almost every kind of shell, Iron pyrites, barytes, Calcarious spars, Cupids wings, green sand,

Lobsters, cray fish, crabs, teeth, fern Echinets, wood, a variety of fossil shells, gupsum selenite Chalk, teeth, palates, vertebral colums of pentacrenite four kinds of Echinetes, a greate variety of shells madrepore's, Alceoniums, Terabratulae Pectens … Alluvium, tusks of the mammoth, teeth of the Rehinocerus, teeth of a species of Bullock, flints Chalcedony agates, jaspers, wood echinetes, alceonium a variety of shells Amonites.' What a list. What a museum.

'I seem to have been only like a boy playing on the seashore,' Isaac Newton famously wrote, 'and diverting myself in now and then finding a smoother pebble or a prettier shell than ordinary, whilst the great ocean of truth lay all undiscovered before me.'

I think of a skinny Dorset girl on an ancient seashore. Mary Anning looked at it differently. The pebbles were not a diversion. She was not there to play. The pebbles, the rocks, the restless layers of shale and limestone, 190, 200 million years old, these *were* the truth – a great ocean of truth, written in stone.

It's hard to say with any certainty whether the rockpools of our coasts have changed – become emptier, less wild, less *fun* – since I was little. Each reach of intertidal seashore has its own ecology, its own rhythms. These are dynamic habitats. It seems likely that, between them, over the last 30-odd years, sewage overflows, rising human footfall, a warming climate and the construction of new sea defences have had *some* impact on these places (these crablands, this kelp country, these winklespheres and limpetaria) – but we don't know exactly what. And the hordes of Victorian amateurs ransacking the rockpools are long-since dispersed, of course. In their place are new threats (there are always new threats). Intensive trawling disturbs offshore mollusc populations, which in turns distorts the connecting webs of species on the fringes of the land. Winkles and cockles are gathered illegally in huge quantities (a problem worsened by the shutdown of legitimate shellfish markets under Covid-19 restrictions).

In any case, Genevieve and Danny will, when they're my age (what a thought!), remember a different seaside to the one I remember. The coast, like the song says, is always changing. The writer Michael Bywater has an entry for the seaside in his book of *Lost Worlds* ('and then it was time to go and it's all gone now anyway, or is it that we've all gone now...') – our seasides, wherever they are, are in so many ways places made as much of memory as of rock and sand, salt and seawater.

Touch is not, for me, the most evocative of senses (that would be smell – and within that, the smell of the seaside, kelp, fish, wet sand, deep-fat fryers, is the most evocative of all). But touch is I think the most persistent, the most prone to haunting, to lingering phantoms. The feel of a cricket ball you still feel hard and heavy in your palm though you've not played in years. Cold sand running liquid between your toes as you stand in the breakers. The feel of barnacled rocks underfoot. The small strength of a hermit crab held in the hand. A limpetshell, translucent in sunshine, sitting like a hat on the tip of your finger. If you remember how it feels, doesn't it mean that somehow you still feel it?

I'm lucky, I know, to have seaside memories at all. Our children are even luckier: grandparents who live *actually at the seaside* – what a thing that would have seemed to me at five, six, seven, hell, twenty, 30, 44 years old. Wakefield, the city where I grew up – the city my dad's from – is slap-bang in the middle of England (strange, then, that my dad has such a penchant for standing on headlands and gazing out at the sea like a sailor's widow in a folk song). We'd take that fortnight every August: me, my brother, my parents, my mum's parents and my auntie. I was lucky.

John Krish's 1961 documentary *They Took Us to the Sea* documents an NSPCC outing to Weston-super-Mare for underprivileged and abused children from inner-city Birmingham. Many of them had never travelled beyond the city's limits. 'Never seen the sea before,' says a boy's voice, off-camera. 'It's got a funny smell to it.' In 2014, Krish told local press that he remained haunted by what he'd seen ('and smelled') when working with NSPCC inspectors in preparation for the film. It's a film about joy, of sorts – but it's really about being allowed to feel joy, about the sadness and fear that can persist when a child's capacity for joy is neglected, left undeveloped. You can see it in their faces, some of them. For all the donkey rides, candyfloss, spins on the carousel. They don't know what to *do* with it.

Years ago, I saw a different documentary about these NSPCC outings. I don't remember all that much about it but there was one snatch of footage that has stayed with me. Black-and-white film, dark and spotty (cheaply shot, or poorly kept since). A high angle, I guess from a prom or cliff path. Children, perhaps two dozen of them, scampering like dark little imps or elves, suddenly *burst* out on to the open sands, and you can see in their movements a sort of ecstatic madness. The sky! The space! The air! These kids, too, don't know quite what to do, except that they know they want to run, and roll, and fall over, and spin, jump, skip, run some more, fall some more, spin some more – so that's what they do. There's no sound on the footage but you can imagine the uproar over the noise of wind and sea. This is about joy, too: literally, joy unconfined.

At Cullercoats, just a short way down the coast from St Mary's and Prampa's Lighthouse, there's a little bay tucked tight against the cliffs, confined by two tall sea walls. The sandy beach has a shallow slope. At low tide the wet intertidal zone is spotted from one side to the other with the casts of blow lugworms (a coiled worm of spent sand, and an inch or two away a little dimple with a hole at the bottom, an analogue of the worm's digestive system). At the north end the seawater sometimes comes rushing in black with sea-coal deposits. We have our beach days here, the four of us – this is where we set up our little beach tent, embark on great undertakings of spade and bucket, kick a football, eat sandwiches, drink from our flask of tea, find unremarkable scraps of seaweed and shout LOOK, LOOK, see globe jellyfish bobbing in the shallows and go *AIEEEE!* Sometimes we walk on the sea walls. Sometimes we have a scramble about in the rich rockpools either side of the bay. We watch the swimmers, the paddlers, the wakeboarders.

Just above us, behind us, on the edge of the land, is the Dove Marine Laboratory. It's a fine, tall building of salt-scoured red brick, built in 1908 on the site of the old saltwater baths to replace a smaller, wooden facility that had been set up there by the Northumberland Sea Fisheries in 1897 but that burned down in 1904. It was funded by the geologist and fossil collector Wilfred Hudleston and named 'Dove' in honour of one of his ancestors, Eleanor Dove. Today, it's part of Newcastle University. Researchers work on deep-sea food webs, marine biotech, sea mammal biology, water treatment, plankton ecology, oceanic climate change impacts. Since 1968 it's been home to the Planktonic Dove Time Series: each month, a seawater sample is taken from the North Sea at a point around 54 metres in depth, about five miles out from the port of Blyth. The sample is fixed in formaldehyde and prepared for long-term preservation. The resulting sample set (that must be, what, 640-odd samples?) is a window on half a century of zooplanktonic and phytoplanktonic change – that is, environmental change. Catherine moved up here when she was little, in about 1982. Some fancy restaurants can bring you a glass of port or cognac that was bottled in the month of your birth – somewhere in the Dove labs, in deep storage, there's the planktonic equivalent. Perhaps they could give us a small phial of '82 vintage North Sea, from that first time she breathed the air here – or the first time I came here, perhaps, or the month of Danny's first paddle, Genevieve's first sandcastle. I love the idea of these snapshots, these moments held static in time that – like ice cores or pollen samples – tell us not only about where we were, where we've come from, but how we got here – how change, that lone constant, actually *works*.

In addition to its research and teaching work, the Dove lab runs outreach activities for schools – Key Stage 1 kids, that's kids between about five and seven, can learn how to touch and hold marine animals, how to treat them gently; they can learn about how and where they live, what they eat (and what eats them), how pollution and climate change can threaten their lives. It sounds brilliant. There's a summer school, too – I'd sign up today, if I were nine (or if they admitted 44-year-old men).

My kids won't go to school around here (they'll most likely go to school around 75 miles from the nearest seashore). They do, though, already have a route into marine biology; they've already learned more than I ever did about whale sharks, yeti crabs, vampire squids, remipedes, sailfish, blobfish, flying fish, walruses, sea cucumbers, comb jellies, slime eels, torpedo rays, lanternsharks and marine iguanas; they know about artificial reefs and kelp forests, algal blooms and whale sonar. They have studied intently and for many months under Dr Shellington of the Octonauts.

Shellington is an amiable sea otter who – in *Octonauts*, a CBeebies television series based on books by Vicki Wong and Michael Murphy – serves as a marine biologist with the crew of the Octopod, a sort of mobile, many-legged under-sea equivalent of the Dove lab (actually, it's never established in the series where Shellington earned his PhD, but he has a Scottish accent, so it wouldn't be too big a leap for him to have studied at Cullercoats). The three chief Octonauts are Captain Barnacles (a polar bear), Kwazii (a cat, who used to be a pirate) and Peso (a timid but courageous penguin, who is employed as a medic). Each episode begins with the Octonaut mission statement: Explore, Rescue, Protect!

Octonauts has been an immense global success since its television launch in 2010, and quite right too: it's funny, gripping, smart and relentlessly – but covertly – educational (my main qualm is that Captain Barnacles appears to lose his Yorkshire accent as the series progress, raising questions about regional bias and equality of opportunity in the Octonaut command hierarchy). It's Genevieve's favourite programme, by some distance; it may be her first great love.

Television, when I was three or four, was less action-packed. Hartley Hare got into the odd scrape with his uncle, there was that time the window cleaner got stuck on the roof in Trumpton, but mostly lives were quiet, and people were kind, in *King Rollo*, *Bagpuss*, *Bod*, *Postman Pat* and the rest. I don't think I learned very much from them, either – except, perhaps, how to live a quiet life and be kind (there are worse lessons). Nothing came along like *Octonauts* till I was a fair bit older – but when it did, I, too, fell in love. I found the values of the Octonauts, translocated to the 1950s and given an American flavour and a profit motive, in the adventures of Hal and Roger Hunt.

It all came back to me when, in the early hours of some weekday morning, with Genevieve and Danny clutching their milk cups beside me on the sofa, I was watching Captain Barnacles struggle with a giant clam that had got hold of his paw. *What Barnacles needs to do here*, I thought to myself, *is somehow sever the hinge of the clam with a trusty knife*. This is not, in fact, what Barnacles ended up doing (he used a length of driftwood to lever it open) – but it was what Hal Hunt did.

Hunt was seized by a clam about the ankle and abandoned by his treacherous rival S.K. 'Skink' Inkham somewhere, I forget where, in the South Pacific. It was not an atypical thing to happen to him, as in the course of the *Adventure* series of novels he faced down a leopard cult, a volcanologist with a brain trauma, a slave-trading sheikh, a judge who is secretly a poacher, an enraged gorilla, a volcanic explosion, man-eating lions, a deranged oceanologist and a murderous business rival. The Hunts – eighteen-year-old Hal, thirteen-year-old Roger and their father John – are New York animal collectors, created in the late 1940s by Willard Price. Price was a reporter, a traveller and (possibly) a spy; his aim in writing the *Adventure* books, he once said, was to 'lead [young people] to read by making reading exciting and full of adventure. At the same time, I want to inspire an interest in wild animals and their behaviour.' Swap 'reading' for 'watching CBeebies' and this is exactly what *Octonauts* does.

The Hunts – usually just the boys, as John tends to absent himself on business matters or because he has a bad back – travel the globe in order to gather up exotic animals for US zoos. The short-term objectives can at times seem a little rapacious ('I can see the Cincinnati Zoo paying upward of $5,000 for this fierce fellow!' one of the brothers might cry), but the broader ethos is – in later books, anyway – environmental. In other ways the books are rather, clears throat awkwardly, shuffles feet, *of their time* – the critic Sam Leith singles out colonialist racism (particularly in the early books), violence and what he rightly calls an 'intensely homosocial' environment – there are no women visible in Hal and Roger's lives (at which point it's worth noting that Barnacles, Kwazii and Peso take the lead in every *Octonauts* adventure, and there are only two women, the mechanic Tweak and the computer operative Dashi, in the six-strong Octopod crew).

Leith also points out that in Anthony McGowan's reboot of the *Adventure* novels, begun in 2012 with *Leopard Adventure*, the Hunts, Frazer and Amazon (a girl, of all things), are not collectors, but environmentalists. I think, when I was seven, eight, nine, I liked that *my* Hunts were collectors. I liked that they got something to take home at the end of it. I felt real satisfaction every time the door of an improvised bamboo cage was finally shut: *got him*. And of course I loved zoos – zoos were *goodies*, zoos were Gerald Durrell and Johnny Morris and David Attenborough (Attenborough's first television series, remember, was *Zoo Quest*, in which he and a London Zoo employee travelled to the tropics to collect animals), zoos were trips to London with my auntie, the elephant house, the penguin pool (we also once went to a zoo in Norfolk, which had a sad, dirty bear – people were throwing him salted peanuts, so at least I had a sense that, even if I was broadly and in principle pro-zoo, not *all* zoos were good).

Most of all I liked that the Hunts were hands-on. This urge again, to touch, to grab, to hold – it's childish, fundamentally childish, and so I loved all that when I was a child. I know Anthony McGowan a little bit, so I asked him how he'd characterise the view of nature articulated in the original *Adventure*

books: 'The natural world,' he told me, 'always remains first and foremost a playground for the Hunt boys.'

It wasn't that I wanted to tussle hand-to-claw with a cheetah (Roger Hunt does this, and prevails by imitating the real-life naturalist Carl Akeley, who in 1896 killed a leopard by forcing his arm *into* its throat; don't worry, Roger only weakens the cheetah and later adopts it as a pet). It was more, I think, that I simply wanted to get in and among it all – I didn't want thrills and adventure, but I thought it would be fun to hold a big snake or have a monkey sit on your shoulder or be friends with a rhinoceros or share your house with a red panda. The point is, I think, that I couldn't see any good reason why anyone *shouldn't* do all these things.

You grow up and you learn a bit more about wildlife, and ecology, and people. You learn that we do have to keep our distance, walk softly, tread lightly among wild things. You learn that as a species we're not unlike Lennie in *Of Mice and Men* – we want to hold the mouse, but we've not quite figured out how to hold the mouse without killing the mouse.

Octonauts, I think, understands this. The Octonauts aren't collectors, of course – that wouldn't make any sense at all, why would a polar bear and a penguin work for a zoo? – but there's something about what they do that has, for me, the same appeal as the *Adventure* books. It's not Explore! Rescue! Protect! – that's all well and good, of course – but what the Octonauts *really* do is deal with all these creatures on their own terms. They don't do anything at arm's length; the animals aren't there to *look* at, they're there to live with, and that means, well, getting in and among them: chasing and being chased, pushing, pulling, helping, racing, bargaining, making friends, arguing, trading, persuading. Every creature is shown respect and care, but none are seen as *precious*. The Octonauts are allowed to live in the animals' world, no strings attached – that's the great trick the show pulls. It isn't for the most part a conservation show – conservation needs a wide angle, or a deep focus, and what we have here isn't that. It's a show about decency, shown in close-up. I call it a 'trick' (and mean it kindly) because sea otters aren't biologists and penguins aren't medics. *Octonauts* works its magic by showing us the best of humankind – because we *do* explore, rescue and protect, sometimes, when we feel like it – and teaming it, in a way that's simply at odds with what humans are, what humans *do*, with something fundamental to wild things: freedom.

One day not far off, when we're at the beach, I'll interrupt the sandcastling or the sandwich-eating and explain to Genevieve and Danny about what they do in that building up there, the red one, the one with the gull on the roof going *pyew, pyew*. It's full of scientists, I'll say. Scientists like Mummy? they'll say, and I'll say no, scientists like Shellington.

They might lose their minds. Or Genevieve, with her grown-up head on, might look at me seriously and say but Daddy are there Shellingtons in *real life*? and I might have to say, well, no, not exactly, these ones aren't otters, and they aren't Octonauts, not really, but look – and here I'll explain as best I can about the deep-sea food webs and the marine biotech and the sea mammal biology and the planktonic ecology (we'll look up phytoplankton and zooplankton on Wikipedia – '*For the character, see Sheldon J. Plankton*' – and YouTube dugongs and sealions). I'll say it's not quite Shellington, but it's still pretty cool, isn't it?

And she might say yes, or she might just shrug and go back to her sandcastle or her sandwich. It won't really matter because after lunch or after the tide's come in over the castle, we'll head back to the rockpools, all of us, and Danny'll still go *look look Daddy* when there's a sea anemone or a starfish and Genevieve'll go *what's in this one Mummy come on Mummy come on* and if I pick up a hermit crab in a winkle shell, little claws rowing at the air, mildly signalling, they'll still go *what is it what is it* and then they'll go *aiee* or *oohh* and then they'll go *put it on my hand Daddy can you put it on my hand no me first Daddy I was first put it on my hand.*

CHAPTER 5

Beech leaf

Hirst Wood, all the time

GLITCH: Grimbots, have you definitely brought me
to the right pristine forest?

GRIMBOT: [affirmative beep]

GLITCH: Gasp. Just look at it! This place is FILTHY.
And the trees are full of holes.

'Endangered Species: São Paulo, Brazil',
Go Jetters, 16 September 2020

I walk to school the usual way: up to the top of our hill, along Peel Street to the residential crescent of Hallcroft Drive and across Cluntergate. It's about a mile. I must have walked this route, there and back, more than – wait while I run this through the calculator – 2,200 times.

Today I'm walking it for the first time in 33 years. I thought it would have changed more than it has – then I could have made a lot of insightful and moving remarks about life and growth and change and the transience of all our childhoods. Really it hasn't changed at all. There are still hot-pink carillons of fuchsia flowers overhanging every other garden wall; there's still buddleia, there's still lilac – these three, all pinks and mauves and purples, were the sacred floral trinity of 1980s suburbia (though I think there's less lilac than there used to be: there's some, and when I smell it it reminds how much more of it there used to be). There are a handful of new houses, by which I mean houses I remember being built. There are still house martins overhead – the eaves of those 'new' houses might have seen 30-odd generations of martins come and go by now. One thing is completely new: along Hallcroft, four or five houses have a pristine sward of fake grass out front. It's hyper-green, unnecessarily, violently green. I can't imagine how it looks in the colour-saturated vision of birds and pollinating insects (but then, there aren't many of those around). Aside from anything else, I'll never quite get used to the sight of people hoovering their lawns.

The entrance on Cluntergate is closed these days, so I walk down to the bypass and along to the main gate. The building is the same – red brick, single storey, infants' school at one end, junior school at the other. It's called an academy now but when I went there it was called Northfield Lane.

The caretaker, a friendly and fit-looking fellow called Gary, meets me at reception (when I was here the caretaker was an elderly man with a bushy grey beard: he was probably about 45). We walk across the car park, which was a playground 30 years ago, and through a gap in a wall on to a stretch of sun-parched grass, where we pause for a minute so I can get my bearings. There are three concrete tennis courts, and a concrete play area with hopscotch grids chalked on the floor; there's a quiet study area in the shadow of a big beech, with bird feeders and bug hotels; there's a little Astro football pitch (somewhat less lush than the Hallcroft lawns); there are allotments, crowded with summer greens in raised beds; there's a big square of bare green, maybe 70 metres across, where we used to hold Sports Day. And there's a little wood, overhanging the top fence.

'I remember,' I say, relishing the moment, 'when all this was fields.'

Gary smiles and leads the way up past the allotments to a gate in the fence.

Then, for the first time in my life, I'm in a wood – what the hell, let's call it a forest – that I planted. That is, I planted part of it. I planted one tree. I think I was nine. Now the tree is 34 years old and towers overhead.

There was nothing but an open grass field and a hem of unmanaged scrub out here when I was at school, when I was six, seven, eight. At nine I was in the third year of junior school, Mr Webster's class (the year groups here now are named after animals: Capybaras, Snow Leopards, Macaws, Hyenas, Mammoths, Wombats, Buffalos, Rhinos, Ladybirds, Meerkats, Chipmunks and Kangaroos – a strange salad of species, but it's surely a welcome string to any teacher's bow to be listed on the school website as 'Armadillos and Maths Lead'). One day we all trooped out to the top of the familiar field – *today, children, we are going to plant trees!*

We'd had lessons about it, I'm sure. We must have looked at those 'life of a tree' pictures – the big diagram of the oak (it was always an oak) with cut-outs showing all the living things the oak supports. We must have learned about the importance of trees. I don't really remember. I remember being given my sapling – I don't remember what it was – and being shown how to dig the hole the right way and put the roots in the earth the right way.

This kind of education was *huge* a century ago. A US 'popular science' magazine from the 1890s reported – not without alarm – that 'in many places in Europe school grounds are very much better managed than in this country ... [school authorities] appreciate the training which results from pruning, budding and grafting trees, plowing, hoeing and fertilising land, hiving bees and raising silk-worms'. A barrage of stats follows: in 1890 there were nearly 8,000 school gardens in Austria; in France, practical gardening was taught in 28,000 primary and elementary schools, each of which had a garden attached to it; in Sweden, 22,000 children received instruction in horticulture and tree-planting; in one province of Russia, 227 schools out of a total of 504 had school gardens; from 1877, every public school in Berlin was regularly supplied with plants for study, 'elementary schools receiving specimens of four different species and secondary schools six'.

We were just twenty-odd little kids getting our school shoes muddy.

Gary shows me the little timber seating area he's built in the wood. There's a totem pole and signs saying what some of the trees are (vandals, he says, have booted a few of them over). It's July, so it's richly overgrown, the clamouring green having lost its spring zip, the juicy zing of sap season, but retaining still half a summer's worth of life and vigour: when I go off the path I'm straight into knee-deep ivy and a grabbing matrix of head-height brambles, tendrils, low-hanging branches and up-leaping beech.

'Do you know which one's yours?' asks Gary.

'It was – it was somewhere in the middle,' I say, a bit help-lessly, forlorn in the undergrowth. I do remember it was in the front row of trees (the wood is about five or six trees deep, eight or nine across). It had a piece of blue tape around it, so I'd know it was mine. I look at the trees around me: now adults, of course, proper grown-ups, veterans of 100-plus seasons.

I am (don't tell Genevieve) terrible with trees. There's an ash – that's not mine, it's too far over. There's a maple, that could be it, and something that might be a lime or might be a wych elm (I look it up later: it's a cherry); next to that there's either a sallow or a mulberry or a hornbeam.

I've often thought birdwatching would be easy if only the damn things would stay still. Well, trees do stay still, they basically do nothing *but* stay still, and I can still hardly tell one from another. Trees are *hard*.

'I'm claiming this one,' I shout to Gary, patting the lime/elm that is in fact a cherry. 'This one's mine.'

It could hardly be more obvious, though, that of course it's *not* mine – that's the point. Even if it had been mine to begin with, it's been everyone else's ever since. All these trees, these oaks, ashes, cherries planted by Year Three in 1987, hang heavily forward, over the fence, over the playing field – they have to be regularly cut back by the council, but they're great in summer, Gary says, for the shade, for the kids, for sitting under, playing under, reading under. This is the deal with trees. We go on, they stay behind.

It's just occurred to me: 1987, the year of the Great Storm, when over the course of two days in mid-October winds topping out at 120 mph ripped across the south of England, and as many as 15 million trees – 15 *million* – were blown down. I remember it from television news reports; it mostly passed us by, up here.

Storms, wrote the great woodland ecologist Oliver Rackham, 'call in question the assumption that the "normal" state of a tree is upright'. The wreckage of '87 represented not an ending but a transformation. The majority of windthrown trees survived: in response to the abrupt change in gravitational circumstances, Rackham noted, most 'sprouted at least from the base, and sometimes all along the trunk'. Those that did die after falling generally did so not from thirst but from lack of sunlight (so shade-tolerant limes, for instance, came back a lot more strongly than, say, light-loving birches). Ancient trees were seldom blown down; 'rotten' or hollow trees proved to be no less resilient than apparently 'healthy' equivalents. Age, decay, collapse: these are not, storms teach us, stages of sad decline, stations of the cross on life's via dolorosa, but components of a dynamic life cycle. Some comfort, I suppose, as we stand in the shade of the trees we planted as children, fingering our grey hairs, thinking about the years (no, the *decades* – Christ).

I notice that there are lengths of cut timber lying in the ivy. Ash, says Gary. Last summer the council had to come and cut down 21 ash trees because they were infected with 'some ash disease' – ash dieback, I guess: the fungus *Chalara fraxinea*, which over the last ten years or so has infected ash trees across the UK. It may be that as many as 80 per cent of our ashes will eventually be killed by the infection (at present the mortality rate of woodland ashes is just short of 70 per cent). I roll over one of the cut logs with my foot and woodlice run for cover. There is more than one sort of storm.

On the way out of the wood, I notice that one of the ash stumps is surrounded by a litter of broken snail shells. It's been used as an 'anvil' – a hard spot where a song thrush prepares its *escargot*. In a wood it's generally a bit too noisy to hear but you'll sometimes catch it in a quiet garden: the *tink, tink* of snails being hammered. I'm glad I found this one.

There are few more facile truisms than 'life goes on'. Life goes on, but there are no ash trees in it. Life goes on, but there are no elms, or whitebeam, or juniper, or woolly willow; life goes on, but there are no jungles, no forests, no trees at all, save oil palms and plantation pines. Life goes on! Except that there are no wildflowers and no butterflies, moths, hoverflies, bumblebees, there are no sand-eels and no seabirds, there are no peat bogs and no clean streams so no otters, water voles, kingfishers, salmon, crayfish, water-crowfoots or starworts, no sphagnums or sundews, no orchids, plovers, greenshanks, hen harriers, not much of anything really, but still – life goes on! Consider the slime mould. Think of the microbe. Life goes on.

I'm spiralling. It happens (it happens more the older I get). 'Life goes on'. It doesn't mean much – but on the other hand, it means everything. It isn't the *what* of it, it isn't the bare fact, 'life goes on' – that does no one any good; it's the *how* of it. How life happens. How life persists. The fallen tree recalibrates, finds its balance and grows again. We mustn't mistake every change for an ending (but then nor must we mistake endings for change: some things do break, some things do stop, some things do have to be fixed). It's this *how* life goes on that means everything. It means everything, that the life of the wood persists. It means everything to walk through a wood and find a dead ash stump and a litter of broken shells.

———

At the bottom of the school grounds, overshadowing the bird feeders and the bug hotel, stand two beech trees. They were there when I was a kid. They were already pretty old then – at a rough guess they're about 190 centimetres across at chest height, and if you multiply that by 0.45 you get a ballpark figure for the trees' age: 85.5 years. The first school, Horbury New Council School, was built here in 1913, 110 years ago. Maybe the children planted the beeches (maybe some of them came back as adults, decades later, to look at them, and stood in the shelter of their broad copper canopy and felt terribly, terribly old).

Beech trees routinely live for 200 years. If they are pollarded – that is, severely cropped to encourage new growth – then 350 years is possible; in upland habitats, they can reach 400. There's one in Surrey that's thought to be 453. There's one in the Italian Alps that may be 520, and one in Maglizh, Bulgaria, that's about the same age. The dodo wasn't yet extinct when these trees germinated. When these trees were saplings there were still aurochs, hulking ancestral cattle, six feet at the shoulder, two feet across the horns, roaming the woodlands of Eastern Europe.

Old beeches are categorised as *notable* (75–150ish years old), *veteran* (125–200ish) and *ancient* (225+) – by these lights, the old school beeches are barely out of adolescence.

Our woods, the woods we walk to, a couple of miles from our house, are beech woods – beech, and oak, and birch. They're ancient woods that sprouted on a heap of glacial till some time after the ice sheets receded. They are – and this is important – close to both a coffee shop and a playground.

Though of course the woods *are* our playground. We run about. We balance on low branches. We pick up sticks (we hit things with the sticks). We hide from dogs. We fall over in the leaf mould. We look under stones. We find frogs. We tramp through mud. We go *ugh* at slugs. We tread in dogshit (we get wiped up). We sit on logs to eat sandwiches. We see jays and woodpeckers. We run after squirrels. We touch holly and go *ow* and then touch it again. We roll down slopes, sometimes on purpose. We touch bark. We dig for treasure. We hear nuthatches whistling (Daddy whistles back, to annoy them, he thinks it's funny). We go inside hollow trees. We find faces in things. We watch for aeroplanes overhead. We find ferns and pretend they're wings. We pick up acorns. We pick up feathers. We pick up beechmasts. We pick up beech leaves.

Of course, saying *the woods are our playground* doesn't wash with a two-year-old and a three-year-old. We still have to go to the real playground afterwards.

But these beech leaves. Beech trees are, we've seen, resilient. They stick around. They get old, they get gnarled and knotted; insects and grubs and birds make holes in them, get under their bark, into their hollows. Great spotted woodpeckers raise a great drumming from the resonant deadwood (we've watched them in our woods, in spring, fighting over the big beeches). Read up on the caterpillars that feed on beech and be reminded of the folk-poetry of moth names: *barred hook-tip / clay triple-lines / olive crescent*. Beeches are homes. More than that, they are communities.

My friend David Haskell is one of the great modern writers on trees, on forests. 'A tree,' he writes in his book *The Songs of*

Trees, 'is a being that catalyzes and regulates conversations in and around its body. Death ends the active management of these connections ... But a tree never fully controlled these connections; in life the tree was only one part of its network. Death decenters the tree's life but does not end it.'

You do not easily get rid of a tree. A tree that dies in a temperate forest – a forest like ours, so a beech, an oak, a lime, a cherry – may lie there dead for as long as it stood there alive. Tree life is stubborn life.

The beech expresses resilience, or perhaps stubbornness, in other ways, too. Across the street from our house there's a long beech hedge that borders a college building. Each autumn its leaves turn red-gold. I wait for them to fall, and they don't: they cling on. Through the winter the beech hedge looks like a wall of cornflakes. This is marcescence.

The mechanical explanation, the *how* of marcescence, is straightforward. The beech takes a lot longer than most other trees to build what's called an abscission layer between the stem of each leaf and the branch to which it is attached. The abscission layer, as it develops, cuts off the leaf from water and nutrition – and so the leaf dies and falls. What's left is a sealed wound, a sort of scab. In most trees, the abscission layer forms in autumn, the leaves drop and the tree is safely shuttered for the winter. So, why does the beech do things differently?

The beech is, of course, doing what's good for the beech. This means – among other things – keeping up with the other beeches. These are big, broad trees with wide, sun-guzzling canopies; living a photosynthetic life in the shadow

of a mature beech is no easy business, but of course, young beeches don't usually have the chance to live anywhere else (the beechmast seldom falling far from the tree), and it's in younger beeches that marcescence is most pronounced. Ecologists suspect that by retaining its leaves till spring the growing tree is, in effect, mulching itself. Leaves dropped in autumn may have delivered the greater part of their nutrient load into the soil by the time spring comes around and that nutrient boost is really needed; a spring drop, on the other hand, might give the sapling that little kick it needs to push on and get some serious growing done before, once again, the green roof closes in overhead.

We're talking fine margins here, though. Beech leaves, like oak leaves, are waxy, and don't decompose quickly – a beech leaf won't decay completely away till around two years after it falls from its tree. Again: resilience, or stubbornness.

You can't cross the same river twice, the saying goes. In the same way, you can't walk (or run, jump, skip, toddle, hopscotch, scoot) through the same forest twice. By the day, by the hour, by the minute, by the second, it's changing, *everything* is changing – and though this spring will look quite like last spring, and next spring will look pretty much like this spring, the reality is that woods are remade – no, woods remake themselves – over and again, cycling through countless regenerations within the span of a human lifetime.

And yet there are these small stillnesses within the churn. The beech leaf that first didn't want to fall and then didn't want to rot – I like this. As I write, it's just over two years since Daniel was born. I like that a beech leaf that fell in Hirst

Wood in his first spring has only now – maybe today, who knows – crumbled finally to dust, to its chemical component parts (nitrogen, carbon, phosphorus, calcium, magnesium, potassium); has returned, finally, reluctantly, to the soil, to begin again.

The US poet Robert Frost, laureate (from the Latin *lau-reātus*, crowned with laurel) of the north-eastern woods, saw well the poetic potential – that is, the potential for complexity, ambiguity, philosophy and perhaps, but only perhaps, resolu-tion – of fallen leaves, and of the marcescence of the beech leaf in particular. Frost's best-known woodland works are about kids swinging from trees ('Birches', 1915: 'Some boy too far from town to learn baseball,/Whose only play was what he found himself') and pathfinding in woodland ('The Road Not Taken', 1916: 'Two roads diverged in a wood, and I—/I took the one less traveled by,/And that has made all the difference'). Both seem to fit this book. Another tree poem, 'After Apple-Picking' (1914), offers one of the best expressions I know of physical memory, of memory that isn't just an old movie, replayed on loop, but is, rather, a thing of the body, a thing of the senses: 'My instep arch not only keeps the ache,/It keeps the pressure of a ladder-round.' This is another thing this book is about (what's more, Genevieve and Danny know a little about apples, too: apples grow on Grandpa's tree in the autumn, and from the back French windows we watch blackbirds bicker over the windfalls).

But beech leaves (I find them in my pockets, in the pushchair sling, in the kids' coat pockets, shorts pockets, hair, socks, ears) are the theme for now.

'What was that / Far in the maples, pale, but not a ghost?'
Frost writes in 'A Boundless Moment' (1923). The poet and his
companion peer into the woods – flowers, they think, spring
flowers, though it's March, not May. They stand together for a
moment, 'in a strange world', until the illusion breaks: 'I said
the truth (and we moved on). / A young beech clinging to its
last year's leaves.' There's more marcescence in 'Reluctance'
(1913), though this is oak, not beech: 'The leaves are all dead
on the ground, / Save those that the oak is keeping / To ravel
them one by one.' And it's here that Frost first engages with
what seems to be, for him, the meaning of fallen leaves:

> Ah, when to the heart of man
> Was it ever less than a treason
> To go with the drift of things,
> To yield with a grace to reason,
> And bow and accept the end
> Of a love or a season?

A decade later, Frost is still watching the fallen leaves, thinking
with them, feeling through them. 'Reluctance' has given way to
'Misgiving': 'All crying, "We will go with you, O Wind!" / The
foliage follow him, leaf and stem.' But the leaves, having taken
flight, seem to be seized by doubt – instead of going with the
wind, they subside into hollows and thickets, gather in drifts
behind sheltering walls. Frost calls it misgiving, a sort of fear,
a havering uncertainty; we might just as well, though, think
of it the other way about – we might think of it as a stubborn
and awkward love of *here*, rather than a quailing fear of *there*.

'I only hope that when I am free,' Frost finishes, 'As they are free to go in quest/Of the knowledge beyond the bounds of life/It may not seem better to me to rest.'

But we needn't all be Hamlet, holding on to life only for fear of something worse ('There are some pitiable people who are flattered when you call them Hamlets,' cries Ivanov in Chekhov's play, 'but to me it's a disgrace!').

I might take the kids to the woods again tomorrow. We might set up camp in the big beech grove, put out water bottles, little tupperwares of rice cakes and sliced peppers on the fallen logs; we might look under stones and wave sticks about and faceplant in the mulch. I might rest on a rock for a minute with a flask of coffee and watch Danny being a knight fighting a dragon or possibly a dragon fighting a knight, watch Genevieve having a sit down ('just having some *quiet time*, Daddy') among the spent bluebells and garlic plants, breathe in the midsummer, listen to the nuthatches.

It's OK to think, *this'll do*. It's OK to want to stick around for a while.

'A mouse took a stroll through the deep dark wood...'

The chances are you'll know this mouse, if you're a parent. You might know it even if you're not. He is the quick-thinking hero of *The Gruffalo*, the children's book that since its publication in 1999 has sold more than 13 million copies. Executive summary: the mouse meets a hungry fox, a hungry owl and a hungry snake, and frightens them all away with a made-up story about the ferocious 'gruffalo' ('he has terrible teeth, and

terrible claws…'). Then – plot twist – he meets a *real* gruffalo.
He manages to talk the gruffalo out of eating him by claiming
to be 'the scariest creature in this wood': when the gruffalo
follows him through the forest, he's amazed to see the snake,
owl and fox fleeing in terror (the gruffalo doesn't realise that
they're running from him, not the mouse). When the mouse
expresses a taste for *gruffalo crumble*, the gruffalo legs it, too.
Then the mouse sits down to enjoy a nice nut.

That it took me 117 words to summarise (and not very well)
the 700 words of the book is testament to the tidy economy of
Julia Donaldson's writing. But there's a tension, a dissonance,
in *The Gruffalo*; from the first page, the first line, something is
amiss. It's less to do with Donaldson's writing than with the
work of Donaldson's co-author, the illustrator Axel Scheffler.
Look at the cover, look at page 1: this, I'm afraid, is no *deep,
dark wood*.

Scheffler's wood is a thriving mixed woodland of both conif-
erous and broad-leaf trees (on the cover, the Gruffalo peeps out
from behind a birch, and among the roots we can see the white-
flecked red caps of *Amanita muscaria* mushrooms, fly agaric,
birchwood specialists). It seems to be spring (the birch is in leaf,
there are flowering primroses), but in almost every picture we
can see the sky – there's no enclosing canopy. A great-spotted
woodpecker drums on what might be a beech or oak. There's
a clean, quick-running stream where a kingfisher watches for
fish and an emperor dragonfly hunts among bulrushes and flag
iris. There are frogs and red squirrels, foxgloves and bluebells,
beetles and weevils, blue and brimstone butterflies. Even the
paths seem to be well-maintained. I don't blame the mouse

at all for wanting to take a stroll here, it looks lovely. But this, Mr Scheffler, is no deep, dark wood.

It's wonderful, of course, that there's such a wealth of wildlife, lovingly observed, splendidly drawn, in a book that's been read by at least – let's check that number again – 13 *million* children, and probably many more (this is a Scheffler speciality: elsewhere in his books we find background sparrow-hawks, bonus magpies, passing herons, incidental pied wagtails, walk-on rabbits). Besides, if it *had* been a deep, dark wood, we wouldn't have been able to see anything: in true wildwood – not that there's much of that left – you can't see much in any direction beyond a few yards or so (even a gruffalo could hide in true wildwood).

It's true that in fairy tales, in folklore, the forest is typi-cally *the dark place* – it's where wolves lurk, and cannibalistic witches build gingerbread houses, and bad things happen to little children who stray from the path. The shadows of the woods haunt us from deep history; our fear of the wood is a primordial fear. And – like all the best fears – it's one that we have learned to relish ('the woods are lovely, dark and deep' goes one of Robert Frost's most famous lines). *This is too scary – no, don't stop!* We watch through our fingers, but we still watch. I think it's a pretty healthy thing in us, as a society, as a species: there's this hulking dark Jungian bear lurking in the cave of our collective unconscious, and what do we do, we creep a little closer for a better look. We poke it with a stick. Come on then. Scare us again.

Kenneth Grahame's *The Wind in the Willows*, a widely loved childhood classic that I cordially loathe for a multitude of reasons, is a book shot through with anxiety, for all its carefree messing-about-in-boats airs, and nowhere more so than in the *Wild Wood*. Timid Mole takes an ill-advised walk there: 'He penetrated to where the light was less, and trees crouched nearer and nearer, and holes made ugly mouths at him on either side. Everything was very still now. The dusk advanced on him steadily, rapidly, gathering in behind and before; and the light seemed to be draining away like flood-water.' Soon, harassed by sinister and malicious beings on all sides, Mole is cowering in a hollow beech, trembling, traumatised: 'He knew it at last, in all its fullness, that dread thing which other little dwellers in field and hedgerow had encountered here, and known as their darkest moment – that thing which the Rat had vainly tried to shield him from – the Terror of the Wild Wood!'

When that kindly Rat realises that Mole is missing, and heads into the Wild Wood after him, he does so only after arming himself with a stout cudgel and 'a brace of pistols'.

I find the Wild Wood very troubling. I don't like to think about Kenneth Grahame's subconscious.

Such woods, though, are not the woods of childhood, of actual, lived childhood. Not many of us grow up near deep, dark woods and those of us that do are seldom tossed in there to fend for ourselves (imagine a British Mowgli, raised by a family of woodlice, mentored by a stern weasel and a free-wheeling beatnik newt). Our true forest archetypes are not Jungian and subconscious; they're not built from what we fear, they're made of what we need – really, they're made from what we want to *do*. You can call them *deep, dark woods* but when we start sketching them, they come out sunlit, spacious, lively – wild, sure, in the sense that they're lawless, ungoverned, untutored, but never impenetrable, never frightening. There are no wolves. There are no ogres. I think the first thing we learn about forests is that they are places to *play*.

If we want to think about forests not as closed-in things but as spaces – if, that is, we want to think less about trees and more about the gaps between the trees – we could do worse than start at 100 Aker Wood.

First, a few technical points, to satisfy the hardcore Milnians (the Pooh pedants, the dedicated Eeyorologists). The Hundred Acre Wood, in A.A. Milne's Winnie-the-Pooh stories, is only a part of what Pooh calls 'the forest'; it's a patch of deciduous woodland, south-east of the Six Pine Trees, north-west of Eeyore's Gloomy Place (Rather Boggy and Sad). Owl lives there, in The Chestnuts, 'an old-world residence of great charm'.

The Hundred Acre Wood was based by Milne – and by Ernest Shepard, Pooh's first and finest illustrator – on Five Hundred Acre Wood, a thick beech wood that formed part of Ashdown Forest in the High Weald of East Sussex. Christopher Robin Milne, A.A.'s son and the original Christopher Robin, used to walk and play there (photographs from the 1920s show a nice-looking little lad with a great hat of hair, posing endlessly with a shabby bear). The main thing about Ashdown Forest, and therefore Pooh's forest, is that it isn't a forest at all. Like the New Forest, which is also not a forest, it was named with the medieval definition of the term 'forest' in mind, as something characterised not by trees but by access, or rather the lack of it: a forest was simply land set aside, typically for hunting, typically by the King (it will be noted that Winnie-the-Pooh upholds this tradition, hunting – presumably in season and with a licence, though it's never specified – for both woozles and heffalumps).

These are mosaic habitats, broad quiltworks of heath, moor, woodland, scrub. It's a setup that in recent years has received a lot of attention from ecologists. 'Most species require a range of elements within a site or a wider landscape in order to complete their life cycle,' runs a Natural England explainer. 'Many of these elements, such as small patches of bare ground, tall flower-rich vegetation, or scattered trees and scrub, are often absent from the English landscape, and even from some of our most important wildlife sites. This has contributed to serious declines in many species, with some now close to extinction.' So a good mosaic offers a bit of everything, wet and dry, sheltered and open, flower-rich and forested and so on.

There are no doubt a decent number of Wols in Ashdown Forest, and no shortage of Rabbits (and his Friends and Relations). There are no Kangas and therefore no Roos. There aren't any Piglets or donkeys, unless some quixotic soul takes them there. There are no bears.

What there are are marsh gentian, bog asphodel, ivy-leaved bellflower, birds-nest orchid, creeping willow; there are nightjar, cuckoo, Dartford warbler, turtle dove, hobby, firecrest, woodcock; there are golden-ringed dragonfly, purple emperor and silver-studded blue butterflies. Because it's a mosaic habitat it's rich in diversity, and because it's rich in diversity it's rich in possibility.

This is what I think kids see in forests. Possibility.

I never read *Winnie-the-Pooh* when I was little. I learned to love it as a teenager, after finding a copy on my grandad's shelf; now I find it a little harder to stomach the whimsy (Dorothy Parker, who loathed Milne's work, dubbed him 'Whimso', and didn't mean it kindly), but Genevieve and Danny have already started on the abridged versions – they'll be full-fledged Poohvians soon enough.

I can think of a few significant trees from the books of my childhood (Enid Blyton's Magic Faraway Tree, the 'huge tree' under which Fantastic Mr Fox and his family live). Given the choice, though, there's only one invented forest I'd really love to go and get lost in – one that gives me the urge, as John Muir once put it, 'to throw a loaf of bread and a pound of tea in an old sack and jump over the back fence'.

The woods in the *Calvin and Hobbes* comic strips are vast, sunlit, full of slopes and precipices, streams and bogs, rocks and trees – they are emphatically woods for *doing*. I have a print of a *Calvin and Hobbes* panel; I ordered it for one of the kids' bedrooms, but it's been propped on a shelf in my office for years. There they are, in Calvin's wagon, rocketing along a forest track, under a clear sky, through a foreground of boulders and birch – *there's never enough time*, says Calvin, *to do all the nothing you want*. Woods for doing, then, even if what you're doing is nothing much.

Bill Watterson, who created Calvin and Hobbes and wrote and drew them for a decade, from 1985 to 1995, has said that he grew up playing in the woods that backed on to his childhood home in Cuyahoga County, Ohio, but that his woods were a 'brambly swamp' – Calvin's woods look more like a national park. Again, if you're drawing a forest for kids, you'll need to draw trees, yes, but more than that, you're going to need *space*.

I came across *Calvin and Hobbes* in the late 1980s, I think, when it started being syndicated in my parents' *Daily Express*. A chaotic American six-year-old and his sentient toy tiger, with various recurring themes and characters: that was about it. I can't remember what I made of it at first; I know that I came to love it.

Watterson never licensed *Calvin and Hobbes*, expressing instead his commitment to artistic integrity – there were never *Calvin and Hobbes* T-shirts, or bedspreads, or plush toys, pencil cases, rucksacks, car stickers, tote bags, coffee mugs, anything. It was a bold and unusual move, and a genuinely heroic stance to take against commercialisation (in the United States of the early 1990s, at that). Watterson struggled for years with the Universal Press Syndicate for creative control over the characters he created, and I'm really grateful that he did. *Calvin and Hobbes* is special to me in part because – even though I *know* it's been syndicated globally for decades, it's all over the internet, *everyone* has read it – it still, in some curious way, feels like mine, my own thing. If I could buy a furry Hobbes keyring or a talking Calvin doll, I don't think I'd be able to feel the same way.

This again. Our things; the things that are everybody's, the things we feel are ours.

Calvin and Hobbes was, among so much else, an environmentally minded cartoon from the get-go (though I doubt anyone at the *Express* noticed or cared). A 1987 strip shows Calvin discovering that part of 'his' forest has been cleared to build fancy apartments. The storyline ran for three days. 'It took hundreds of years for these woods to grow, and they leveled it in a week,' Calvin says, mooching through the chainsawed stumps. 'Eventually there won't be a nice spot left anywhere.' Then the payoff. Calvin: 'I wonder if you can refuse to inherit the earth.' Hobbes: 'I think if you're born, it's too late.'

Watterson was still thinking of the woods – and of human stupidity, human recklessness – in the mid-1990s. 'I was reading about how countless species are being pushed towards extinction by man's destruction of forests,' Calvin says to Hobbes as they stroll through their woods in a 1994 strip. 'Sometimes I think the surest sign that intelligent life exists elsewhere in the universe is that none of it has tried to contact us.'

If you're not familiar with *Calvin and Hobbes* it might not at this point sound like a laugh riot. There's a storyline from the 1980s where Calvin takes home an injured baby raccoon and it dies, and Calvin and Hobbes lie together in bed and ponder the Big Questions ('It's either mean or it's arbitrary but either way I've got the heebie-jeebies'); there's a weekend strip where they find a dead bird, and talk about the meaning of life ('I suppose it will all make sense when we grow up'; 'No doubt').

But then, there's also a strip where nothing happens except that Calvin pokes some mud with a stick ('ewwww'), pokes it some more ('ewwwwww') and then joyfully plunges knee-deep into it ('EWWWWWWW!'). Calvin contains multitudes.

I might order a print of that one for Daniel. It's very him.

———

When I was a kid there were two kinds of wood near us. One was the kind of wood that didn't really have a name, just a sort of geographical gesture, like the woods around Addingford, never called Addingford, always called 'down Addy', and not, at that time anyway, a real place, a village or a hamlet, but a sort of reclaimed light-industrial zone, all sheds and wood warehouses and dilapidated bridges, peeling paint and barking dogs, a place between the canal and the Calder and the railway junction. Nowadays you'd call these 'edgelands'; Richard Mabey called them 'the Unofficial Countryside', the hedges of nettle and rosebay under the humming power lines, the swallows hunting over the rank grass of the pony's field. Bindweed and motherdie and BMX tracks in the mud. A kid I was at school with used to take his ferrets rabbiting down there.

Perhaps you could call it a feral landscape – returned some way to the wild, but by no means all the way; overgrown, green with moss and rot, but still a human place.

The woods here weren't dark and deep, but they were certainly creepy, not because they were 'wild' in the usual sense, not because there wasn't anybody there, but because there might be *some*body there. Junkies, murderers. People hiding dead bodies. Zombies, robbers, madmen (I'd heard the one about the escaped killer banging the wife's severed head – *thunk … thunk … thunk* – on the roof of the husband's car). It's daft in some ways but not in others. Nothing bad ever happened to us down Addy. Once we saw a man walking stark-bollock-naked down the towpath but that was about the worst of it. Still, my mum didn't like taking me down there by herself – if I'd been a girl, I likely wouldn't have been allowed down there at all. Just too quiet. You say it, think it, with a sort of shrug – and that's terrible. Too quiet, too empty. You don't know what might happen (if you're a lad, probably nothing; if you're a girl – you don't know what). The 21st-century fad for *edgelands* – an understandable fad, given that edgelands are often the only lands we know, the only lands we have – come with this heavy caveat. The human countryside is a place out of balance; what it means (what it says to us, what it stands for, what it promises) is one thing to me and another to a girl out birdwatching or bug-hunting, a woman out looking for wildflowers or fishing in the Calder–Hebble or just walking, just sitting, just whatever. There are many sorts of wilderness.

There were a fair few woods like this around where I grew up and there are a fair few where we live now. You never know when the forest path might abruptly bring you alongside a sheet-metal workshop or a meat processing plant. If the weather's all right and I have Danny and Genevieve for the day I'll take them along Gill Beck Valley; we look for ladybirds (there are loads), feed the ducks, watch for jays and walk the leafy ridge that overlooks Henshaw Timber and Building, Phoenix Coating Solutions, Wrights Recycling Machinery. My favourite birding spot, where I show the kids where a kingfisher was, just a second ago, you just missed it, sorry, and tell them that there's often roe deer here, if you'd just shush for a minute, shush, *shush*, is squeezed into a 50-yard channel between the river and a large manufacturer of radiators, oil coolers, intercoolers and fuel coolers (the air smells of hawthorn and hot metal; the kingfishers don't seem to mind).

A short way along the Leeds–Liverpool canal from our house, heading east, there's a curious stretch of sheds and shanties, pigeon lofts, drunk fencing, small goats, allotments, lost-looking garden furniture, chickens scratching about in shale, home-made 'beware guard dogs' signs. I took Danny for a walk up there when he was a baby, in the sling. It felt very like down Addy. I felt like I might meet my schoolmate Martin, out with his ferrets, carrying a dead rabbit.

So this is one kind of wood: a happenstance wood, likely newish, likely short-lived but who knows, a mess, unmanaged (who would bother?), rich in ivy, midges, robins, woodpigeons, stinking of elder and lime (lime smells 'like a teenager's bedroom', a friend recently remarked). These woods live where we let them live, for as long as we let them live – it would be a few days' work, I'd guess, to grub them out, if we wanted the space (if we ran short of sheet metal or needed more coating solutions).

There is something faintly comic – as well as faintly sinister – about the busy forest, the peopled wood. Terry Pratchett set his fantasy novels in a mock-medieval post-Tolkien Discworld (not in post-industrial West Yorkshire, and I believe they are poorer for it), but he picked up on the same idea. In his book *Lords and Ladies*, a troupe of amateur actors are looking for a quiet place to rehearse their play. 'You'd have thought the Blasted Oak would have been safe,' complains one. 'Half a mile from the nearest path, and damn me if after five minutes you can't move for charcoal burners, hermits, trappers, tree tappers, hunters, trolls, bird-limers, hurdle-makers, swine-herds, truffle hunters, dwarfs, bodgers and suspicious buggers with big coats on. I'm surprised there's room in the forest for the bloody trees.'

Well, this is the thing: there barely is. Shoved-over, pushed aside, squeezed out, still there. Again, we see the resilience of trees. These woods are still woods, just about, more or less.

And then there was the other kind. Not *exactly* the Calvin kind but I'd like to think we all have – we all certainly *should* have – our own-scale Calvin woods. Coxley Valley sounds like a faux-organic farm invented by a supermarket chain to greenwash a line of misshapen vegetables, but it's really just an area of woodland a couple of miles south of where we lived, just far enough out into the actual countryside to take on a different character (in memory, a different *colour*, too – a sun-washed yellow, rather than the dusty dark-greens of ivy and August sycamore).

I've said that woods, for kids, are doing places, and Coxley was a doing place. First it was mostly for running around wildly, and then for playing Army ('nuh-uh-uh-uh-uh-uh-uh-uh, got you, got you', 'no you never, missed, missed', 'no, got you, nuh-uh-uh-uh-uh-uh-uh, got you that time, nuh-uh-uh-uh-uh-uh-uh, pkhoo, pkhoo', 'awwwrgh, pkhoo, pkhoo', just like in the real army); it was for riding BMXs up and down the sun-baked dirt slopes (I called my BMX 'Blue Thunder' and – why, why am I putting this in a book? – did sotto voce commentaries on the races I pretended I was in), and a bit later it was for clattering down timber steps on a mountain bike without any brakes; it was for swinging across Coxley Beck on a length of blue nylon rope (it's always blue nylon rope in these places, I don't know why); it was for legging it

up whatever tree you could leg up – the bark worn to a glossy finish by generations of kids' clambering feet – and seeing if you could drop back down without doing your ankle on a root; it was for sitting in a clearing in the sun in the summer after GCSEs with your mates and that girl you were in love with that year, the one with the indie pigtails and the pink-laced DMs; it was for all that, that was what woods were about, that was what woods were for.

There were also woods – woods nearer home – and a clearing called The Lump, where kids my age, my mates, went to sit around fires and drink cider and smoke weed and get off with each other, and it'd be nice to say I did that, too, but I was a quiet kid when I was that age, and I mostly stayed home in the evenings. Anyway, I'd have to check the statute of limitations before I told you anything more about The Lump.

I don't remember doing much birdwatching at Coxley. I must have – when I was younger, I'm sure I must have. Of course, once puberty kicked in and I twigged that birdwatching was Not Really Cool, I gave the whole thing a rest for a while (I came back to it, not because I realised that being cool didn't matter, but because I realised it was far too late for me to try to be cool in any case). But even when I was young and mustard-keen, I was quite a bad birdwatcher, impatient, distracted, *bored* when things weren't quite as my bird books had told me they were going to be. My first birdwatching memory is of going to Cawthorne – woods again – with my parents and my brother. I remember it was damp and I remember I didn't see anything (where, where were my wrynecks, my golden orioles, my hoopoes, my nightingales?). My brother, who had and still has no interest whatever in birds (or plants or beetles or trees or anything *uncivilised*) lay down in the undergrowth and made up stories about some small people who lived in a spider's web. But my mum tells me that she was crouching quietly a short way away, watching what she swears was a woodcock, when I came crashing through the bracken, shouting: 'HAS ANYBODY SEEN ANYTHING YET?'

I lacked focus, it's fair to say. But it's also fair to say that whether or not I watched birds or hunted bugs in Coxley or Cawthorne, whether I paid any attention at all to anything in any woods, or just mucked about with bikes and ropes and rivers and rocks, is neither here nor there, really, because the point of being in the woods, the *important* thing about being in the woods, is being in the woods.

Epilogue

I'm writing this on the hottest day the UK has ever known.
The news just came through that the temperature at London
Heathrow has topped 40°C – it's not long after lunchtime,
so there's plenty of time for it to climb higher, and not only
in the south-east. Last night, the UK logged its highest ever
overnight temperature, at, of all places, Emley Moor in West
Yorkshire, where it didn't drop below 25.9°C all night. I know
Emley Moor well; I used to cycle up there all the time, up a
variety of testing hills, sweating cobs in the summer sun, when
I was a teenager (why? *Because it was there*).

I just went outside for five minutes. It was all right, for five
minutes. But it's at least 36°C here and rising.

I wasn't born in 1976, when the UK experienced a run
of record temperatures and a fierce summer-long drought,
reservoirs ran dry, heat-related deaths spiked and fires ripped
through heaths and woodland (it wasn't all punk and Jubblies
and Raleigh Choppers). I was around, though, for the record
high temperature of August 1990 (37.1°C), and the record high
temperature of 2003 (38.5°C), and the record high temperature

of 2019 (38.7°C), and I remember the peaks of August 2003 (36.4°C), July 2006 (36.5°C), July 2020 (37.8°C) – in fact, nine of the ten highest temperatures ever recorded in the United Kingdom occurred, not just in my lifetime, but in the past 35 years; seven of the ten occurred since 2000.

I told Genevieve, at the weekend, that it was going to be *really hot* this week. Do you know, I said, it's going to be *the hottest day EVER*.

'Why?' she said. It's what she always says. But of course, it's also the right question to ask.

A large part of being a parent is telling stories, and there are lots of stories to be told here. Of course, there are stories about *baddies*, the oil companies, the profiteers, idiots and denialists in politics and the press, leaders who have been weak or complacent or preoccupied with things that shouldn't matter when the world is on fire, literally on fire, and there's not much time to do anything about it, not to mention most of the rest of us, who feel we're doing our best – we're tired, we're busy, we're broke, give us a break – but aren't doing enough.

But I'm sick of those stories. They're important but I'm sick of them.

I'm going to tell my children stories about energy. I'm going to have to start by explaining what a kilowatt-hour is, how a solar battery works, Betz's Law, the basics of electromechnical generation.

Actually, I might leave the technicalities to Catherine (she does that stuff for a living). I'll tell simpler stories (I do that stuff for a living).

Wakefield, where I was born and grew up, is at the heart of the South Yorkshire coalfield. That should mean, I suppose, that I have some sort of inherited stake in coal, some atavistic investment in the culture of the pit – but in fact, I don't really know much about coal mining at all. Where we lived was outside the mining heartlands, Featherstone, Knottingley, Sharlston, Hemsworth – all places out to the west, where I seldom had any reason to go. I didn't know any miners' sons, miners' daughters. My dad, who grew up in the city centre, remembers the clatter of miners passing his window on the early shift, but the industry barely touched me at all. I remember the strike of 1984–5, but mainly because of the power cuts; I remember Arthur Scargill on the television, and I remember being told off for saying 'scab' (I'd heard it on the *Six O'Clock News*). That was about it.

But I think I know more about it than I would if I'd grown up somewhere else, somewhere beyond the coalfields. The National Coal Mining Museum at Caphouse Colliery was just up the road – the pit gear was still in place, and you could go down the pit, with an ex-miner to show you around, and at one point he'd get everyone to switch off their helmet lamps to find out what proper darkness looks like. At the end you got a piece of coal to take home with you.

Not far along the road from Wakefield to Leeds there used to be two opencast coal mines, St Aidan's and Lowther North, and a colliery, Savile Colliery, at Methley. In March 1988 – I was nine – the River Aire, which flowed right by, broke its banks; water surged into the pit workings, tumbling 200 feet to the bottom in a monumental cataract. Around 17 million litres of river water flooded the site (at one point the adjoining River Calder began to flow backwards as a result, which sounds like some terrible medieval portent). Eventually, the water found its level; the mine was now a lake, with 2.5 million tonnes of coal underneath it.

Ten years on, the miners were back at it – 200 million pounds' worth of pumping and restoration works were completed, and the coal was once again there to be raked from the earth. They got four more years; in 2002, with the English coal industry in a terminal condition, the mine was finally closed. Now St Aidan's, Lowther North and Savile Colliery are something else entirely – an RSPB nature reserve. You can go there and see black-necked grebe, bittern, marsh harrier, peregrine, but the first thing you're liable to see is 'Oddball', a 1,000-ton BE1150-B walking dragline, a surface excavator the size of a house and at one time one of the largest machines on earth. It stands by the entrance to the reserve, a vast monument to the 'sunshine miners', as opencast pitmen were called, and to the history and heritage of coal here. Kestrels nest in it now; starlings throng noisily on its high cables. I've a picture of Danny standing in front of it in his blue splash suit. He looks an inch tall.

So I can be sentimental about coal. I'm from here and it's from here, it was mined by people who talked more or less like me, in an oblique and subtle way it's part of me. I was in a brass band when I was a kid – that wouldn't have happened, there wouldn't have been a brass band without coal (the band used to tour the pubs of the town, playing carols and passing the hat, every Christmas Eve, and I never thought it the slightest bit unusual until, one year in the Cricketer's Arms, Catherine, who was visiting at Christmas for the first time, called her brother and said: 'I'm in a pub and *there's a brass band playing*, HOW YORKSHIRE IS THAT?'). I'm fond of these sorts of understated connections between industry and community – like how in fish and chip shops, no matter how far from the sea, you'll usually see an RNLI collection box.

Coal is dead in this country, and that's good. It's dead but it'll never quite be gone – we've put too much of ourselves into it, invested too heavily, in every sense, to let it go. There's an interview with the great Welsh actor Richard Burton, whose father and brothers were miners, where he talks of his father's *love* of the pit, of the coal seam, not respect, not any sort of grudging accommodation, love: 'He used to talk about it like some men will talk about women, about the beauty of this coal face.' Miners, he says, considered themselves 'the aristocrats of the working class'. It doesn't really matter whether Burton was sentimentalising or not – even if he was, he was expressing something we've all heard, in the street or in the pub or in miners' strike documentaries or news clips from the Durham Miners' Gala, something that I don't think you can get shot of in a couple of generations just by shutting the pits. Stories persist in spite of us.

I'm not sure when I first heard of renewable energy, wind, solar, tide, but I know that when I first started thinking about it, as a kid, my knee-jerk opinion was, well, *duh*. You mean we can *do* that? We can run our fridges and streetlights and televisions on *wind*, literally *air*, and here we are fannying about with coal and oil? Come *on*. Oh, and I had another thought – volcanoes! Why aren't we powering everything by volcanoes? Of course, the very good reason is that we are not Iceland and we do not have any volcanoes (the people of Shropshire would likely object to any plan to reactivate the Wrekin). But still, it all seemed so maddeningly *obvious*. It always has. Honestly, even now I have a better understanding of some of the challenges, some of the downsides, it still does. Why are we not throwing every penny we have at this?

One reason is we haven't been telling the right stories.

There's a concept in Japan, *kojo moe*, which I think translates literally as 'factory love' but describes a 21st-century fashion for industrial nostalgia. Coach trips to derelict automobile plants, glossy photo books of decommissioned power stations, that sort of thing. Drive north over the A19 flyover at night and tell me you don't feel it: the industrial Teesside skyline, underlit, ferrous, monumental, chemical plants, asphalt factories, engineering works, hard things to love in a lot of ways, but powerfully evocative, surely, in the same way as a shipyard or a coalmine, of great undertakings, great resolve, fierce industry, aspiration, ambition, perhaps a terrible sort of ambition, but still.

Maybe it's just me and a coachload of Japanese hipsters who feel that way. The point is, it doesn't have to be nostalgia – these things, these monuments to doing, building, making, working, they needn't be old, we needn't look to the past to see what humanity can do when it bends its back.

The seaside has always been a place of work as well as of play. I like a working seaside: the Fish Quay at North Shields, say, or, on quite a different scale, the great cranes and container ships at Felixstowe (watch them from across the bay as you share your chips with the herring gulls on the shingle beach). Then there are the power stations, at Heysham, at Torness, at Dungeness in Kent (one of the best nature reserves in the south-east sits in the shadow of the nuclear reactor at Dungeness B).

From the lighthouse on St Mary's Island, you can look north and see the town of Blyth. Once it was known for coal and shipbuilding (HMS *Ark Royal* was built there); there were two coal-fired power stations and a thriving industrial port.

The port's still there. The rest is gone: no one's digging coal, no one's building ships, the power stations are long since hauled down, and here, at about this point, is where the songs usually get written, and the nostalgic films get made, and the stories get told, here in this nostalgia sweet spot where things like mines and shipyards are far enough away for them to lose their sharp edges, to dissolve a little in the memory's focus, but near enough, just about, for them still to leave an ache when they go.

But new stories are being written. Seven wind turbines turn on the long arm of Blyth Harbour. Once there were two more offshore, out in the North Sea, the first offshore turbines in the United Kingdom and one of the first in the world; they're gone, but now there are five more, bigger and more effective, a little further out (photographers on Whitley Bay beach complain that they spoil the long shot of the lighthouse – that's one view, but it's not my view). Of course, these turbines are small fry nowadays – now, when there are 174 turbines at work off Hornsea and 165 more on the way, there are 190 on

Dogger Bank, 87 off Walney, 91 just north of Blakeney Point (the sea areas of the Shipping Forecast, Forties, Cromarty, Forth, Tyne and so on, themselves an industrial artefact, are sometimes thought of here as a sort of national poetry, an island nation's secular litany, but try this: Hornsea, Moray, Triton Knoll, Walney, London, Beatrice, the great windfarms of the early 21st century – because we're going to need new poetry). From an environmental point of view – that is, the point of view where I don't want everyone to die in fires or drown in rising seas – I think it's marvellous, joyous, that this is happening, that these things are going up everywhere, that wind (it's literally just *air!*) now provides about a quarter of the UK's electricity. But I love it, too, for the same reason I love the Teesside skyline, the same reason the sight of pithead winding gear makes me feel a bit funny, the same reason I stopped the car once to take a photo of Torness Power Station – *industry*, in its basic sense, means something to us, tells its own story, and I think, where we talk about wind farms, where we talk about solar arrays, or tidal barrages, or hydropower, or any of that, any of those ways forward for us, we haven't said that part loudly enough. The question of what we can *do* is an exciting one.

Nearly 200 turbines, each 100 metres tall, 150 metres in wing-span, out there, impossibly, in the open North Sea, harvesting the trade winds, screwing energy into the earth, where before there was nothing but grey swell and sky – this is a story about doing. We almost left it too late to tell.

My cufflinks have co-ordinates on them: 53°51'24.2"N 1°47'12.4"W. The cufflinks were a present from Catherine on our wedding day; the co-ordinates mark a spot at the top of Baildon Moor, a few miles from our house. It's where I proposed. It's where we've been with the kids, Danny on my back, I think, Genevieve in her wellies, up past the golf course, up to the whitewashed bollard of the trig point ('*I* want to touch it first!' 'No *I* want to touch it first!' 'Well, look, why don't you *both* touch it first?' 'No!' 'NO!'). You get meadow pipit, skylark, snipe, golden plover, wheatear, kestrel up here. In spring or autumn you might get anything. And you can stand at the trig point and look south-west and there in the distance is the windfarm on Ovenden Moor, near Halifax: nine slowly turning turbines, visible, on a clear day, from 35 miles away. I'll hoick the kids up on my shoulders so they can see them better. Tell them some stories: where we've all been, where I hope we're going, what we can all do.

Having kids didn't make me worry about climate change. I was already worried about climate change, because of everyone else's kids. Now, I suppose, it's *personal* in a way it didn't used to be, but the needle on the worry-meter has hardly ticked up at all – there wasn't much room for it to do that (it doesn't go up to eleven).

There's an increasing number of people who are so worried about climate change that they won't have children – either because having children will only make things worse, or because they don't want to bring children into a world that's going through catastrophic warming. Around a quarter of child-free couples cited climate change as a reason why in one recent survey; a Morgan Stanley analysis concluded that climate anxiety is 'impacting fertility rates quicker than any preceding trend'. Maybe if I'd thought about it more I'd have made the same decision. Too late now.

It'll be right is my go-to response to most types of crisis, which is sometimes quite stupid of me and sometimes quite sensible. I won't say it about climate change, because climate change *won't* be right, it's already terrible, the glaciers, the heatwaves, the coral reefs, the sea levels, the wildfires, it's terrible and even in the best-case scenarios it's going to be a long time before it's any better.

And yet it's also true that the world has often, in the past, been terrible, worse than we can imagine, and children have been born anyway, and whether or not things have got *better*, things have, in any case, carried on, and we're here, after all, which seems better than the alternative. And then there's the wind farms and the solar technology and, and, and…

Perhaps this is just us flailing as we fall.

Besides, my kids *will* probably be all right, whatever 'all right' will mean when they're twenty, 40, 60, when I'm gone (this, for me, is the most painful thing to imagine, as a parent: not *being gone*, I'm OK with that in general terms, but the idea of our children getting old, and facing up to everything that comes with getting old, and us not being there to tell them it's all OK: it's something that had never occurred to me until Catherine mentioned it a couple of years ago, and now I think about it every day, *thanks, darling*). They'll be all right because they'll live in a fairly safe country in a more or less safe part of the globe – it's not much to leave them with (it's less than we were left with, which in turn was less than our parents were left with) but it's something.

We need to leave something more, of course. A little hope.
That would be nice.

'I was self-sufficient,' writes Sam Gribley. 'I could travel the
world over, never needing a penny, never asking anything of
anyone. I could cross to Asia in a canoe via the Bering Strait.
I could raft to an island. I could go around the world on the
fruits of the land. I started to run. I got as far as the gorge and
turned back. I wanted to see Dad.'

This is the last chapter of *My Side of the Mountain* by Jean
Craighead George, a book that, until quite recently, I didn't
know I'd read. I should have known, though.

In our woods, Hirst Wood, quite far through, along the fork
that takes you to the river, there's a big tree stump, hollowed
out by rot. I'm not sure what sort of tree it was, or is. It's
about a metre across. It has slugs and fungus. Genevieve and
Danny take turns getting inside it. Sometimes the one who
isn't inside it goes around the back of it and finds a spyhole in
the bark or else reaches up to look over the top of it: *Peeping!*
Then there's a lot of laughing and nonsense and then they
swap over. There'll be sticks involved, fronds of fern, put to
arcane uses. There are often butterflies to watch here, ringlet,
speckled wood. There's a steepling foxglove where there are
always bumblebees. *I could leave them here all day*, I sometimes
think. I could go and get a coffee. They'd still be here when
I got back, shouting at slugs, hitting each other with bracken.

At first the idea put me in mind of a half-forgotten Enid
Blyton book, *The Secret Island*, in which four children escape

a harsh life with an unkind uncle and aunt and go to live on 'a mysterious island, lonely and beautiful', in the middle of a lake. It's a book that's rich in nature: the island is lush with willows, alders, hazels, elders, silver birches and oaks; moorhens nest there, reed warblers sing, the air smells of wild myrtle (Blyton was strongly influenced by the work of Richard Jefferies). I would have loved all that, of course, but the parts I really remember have more to do with what the kids are doing, Harriet, Henry, Nora and Jack, with their adventures, and, especially, their self-sufficiency. Jack, a cheeky, dauntless local lad, of rural working-class stock, knows how to catch rabbits and which fruits and plants they can eat; a willow tree at the edge of the water provides a snug shelter and – I remember this part very vividly – a hollow tree is used to store supplies (I have a very clear mental image, formed in my tiny head at the age of five or six, of Harriet setting out their provisions on exposed roots, as if on pantry shelves).

I was watching Genevieve and Danny and thinking about Harriet's provisions when Sam Gribley came running back to me.

My children probably won't have much of this, this sensation of half-remembering, of things drifting deep in the memory, hard to see, almost out of reach. Not much is lost, nowadays.

I've no idea when I read *My Side of the Mountain* but leafing through it now I know it had a stirring effect on me when I was nine, ten, whatever I was. I must have had it from the local library or from school because if I'd had my own copy

I would have read it over and again and I'd never have forgotten a word of it.

Published in 1959, it tells the story of Sam Gribley, a young boy who runs away from his family in New York and goes to live in a hollow hemlock tree in the Catskill Mountains ('Sure, go try it,' cries his dad, cheerfully. 'Every boy should try it!'). He catches fish with a whittled hook, trains a young peregrine (he calls it 'Frightful', a lovely touch), skins and butchers a deer, gathers mussels and crayfish, improves his cosy hemlock. It's impossible to read the book and not know that, one, the person who wrote it loved wild places and wild things and, two, the person who wrote it knew what they were talking about.

Jean Craighead George grew up in Washington DC in the 1920s and 1930s. Her father was an entomologist, specialising in beetles, but that might suggest a dry, academic interest, pinned specimens, mothballs, cardboard labels, and that wouldn't be right: her whole family were hands-on nature enthusiasts, explorers, collectors ('By the time I got to kindergarten,' she recalled, 'I was surprised to find out I was the only kid with a turkey vulture.'). They'd do their adventuring in the woods around Washington and at the family's summer home at Boiling Springs, Pennsylvania: this place, all creeks and hills and *Calvin and Hobbes* forest, was 'the basis for generations of naturalists', according to Jean's daughter Twig. It was a place for running wild in nature (Jean, reputedly, would kick off her shoes at the start of the summer and not put them back on until it was time to go back to DC).

Jean's brothers, the twins Frank and John, became ecologists, experts first in hawks and owls – they raised a baby owl together as boys – and later in the grizzly bears of Yellowstone. Jean became a journalist and writer, married an ornithologist and amassed 173 household 'pets' (one was an owl that liked to take showers: 'Please Remove Owl After Showering,' read a sign for houseguests). She even wrote one of those badger-under-the-bed books I must surely have read at some point: *The Tarantula in My Purse* (1996).

I never wanted to run away from home when I was a kid, not even once. There was never anything there that I wanted to escape from (Sam Gribley, though he loves his family, has to share a small city apartment with eight brothers and sisters, as well as his mum and dad). This is the thing about running away to the wilderness: however badly you think you want the thing you're running *to* (I wanted to be Sam Gribley, I wanted that snug hemlock stump with its warm fire and deerskin door), I think you need that little push from behind, too, that little nudge in the back, some kind of impetus that has to do with the *from*.

Think of Darwin, the barnacle, again: *I am fixed on the spot where I shall end it*. I'm attached to where I live now, the house, the town, the pubs, the playgrounds, the friends and neighbours and of course our woods, our moors, our fields, the blue tits over the road, the peregrine on the mill chimney – I love all these things. But I'm not *fixed* here. I'll live in other places, I'm sure, in the course of my life; I'll live near the Tyne, at some point, I'm sure – I'll live near the sea – maybe I'll live in Scotland, or in London, or in the north-west somewhere; I've never travelled much (though I once went to Borneo by mistake) but maybe I'll live abroad: Sweden, Spain, who knows, Canada, New Zealand – the point is, I don't consider myself fixed to a place, and I don't think I ever will. What I'm fixed to is my family – my *up*family, and my *down*family.

'When I told your mother where you were,' Sam Gribley's dad tells him at the end of the book, 'she said, "Well, if he doesn't want to come home, then we will bring home to him".' And so, the whole Gribley family, all eleven of them, decamp to the Catskills and set up home – and not just a home, but a *house* (to Sam's disgust).

We leave Sam there, for the time being, but we know he has in the end the best of both worlds – he has family, and he has his mountains, his peregrine, his *wild*. *The Secret Island* ends much the same way: the children's loving parents, aviators thought lost at sea, turn out to have been not lost but only marooned, and return home to rescue Harriet, Henry and Nora – but they, too, get to keep hold of their wild, as Mum and Dad agree to buy the 'secret island' for them (and adopt the ragged-trousered Jack, while they're at it).

These were both books that helped make me who I am, in ways I can barely begin to figure out. The feat of reverse-engineering will probably always be beyond me. They're books about solitude and self-sufficiency but they're both books that after a long walk, a wild ramble, circle back home – circle back to family.

Genevieve starts school next month. I'm not sure how this happened.

We must have studied nature a bit when I was in primary school. Of course, there must have been a 'nature table' – pinecones, beechmasts, sycamore keys, a pigeon's feather someone picked up in the playground. I think there were display cases in the halls, but I remember only three things very specifically: a sea urchin shell, a glossy cowrie and a stuffed mole. We must have been allowed to touch them: I remember, with a degree of clarity that's almost eerie, how each of them felt. The mole especially. She was mounted prone on a small wooden plinth. Moles, like kingfishers and movie stars, are always smaller in real life than you expect. I remember you could run your finger both ways along its fur. My grandma had a velvet-backed brush that was the same.

We must have had *nature walks* – they were probably more or less like the ones led by Miss Pringle in the Molesworth books ('Off for the woods and keep your eyes skinned. Ha-ha – what do we see at once but a little robin! There is no need to burst into tears fotherington-thomas swete tho he be. Nor to buzz a brick at it, molesworth 2.'). I don't remember being *taught* a lot. The other school in our town, a Church of England school that was known to us as St Snobs, had a wildlife garden (*cor, swank*, as Molesworth might have said) – but I don't remember really envying it. We had our open fields. Once we found a young frog in a tree stump; once we found a dead hedgehog (I was a sententious boy and led my friends in a minute's silence for it). Anyway, you can't play football in a wildlife garden, or bulldog, or tig. We were all right with what we had.

It'll be different, I think, for Genevieve and Danny. Nurseries and primary schools push outdoor education much more than they used to. Genevieve's class at nursery do a day of 'forest school' every week – they don little tabards and form a tidy crocodile and walk up to Northcliffe Woods and a guy with a beard teaches them woodcraft and fire-starting and how to slide down a zipwire. It's wonderful. I'm not sure how much they learn about woodland wildlife, but I don't care (anyway, that's *my* job) – I love that they learn how to *be* in the woods, how to explore, how to play, how not to be scared or intimidated, how to *belong* there. I'm reminded again of those kids on that beach, in the old film – those kids who'd never learned, really, how to be outside.

Maybe when they're older, when they're sixteen, they'll study natural history for their school-leaving exams. It's expected that a natural history GCSE will be on offer by 2025; according to the Department for Education, the qualification 'will enable young people to explore the world by learning about organisms and environments, environmental and sustainability issues, and gain a deeper knowledge of the natural world around them'. There'll also be a focus on the kinds of professional skills that might be needed for a career in ecology or a related discipline: observation, description, recording and analysis.

I found similar language in a set of old pamphlets produced in Indiana in 1898 to help teachers lead nature classes in rural schools: the intention was to bring the children 'into close sympathy with nature, to cultivate the habit of accurate observation, and to give them such information concerning these matters as will be not merely of present interest and use, but which will serve to give added power and usefulness to them in after years, whether they remain upon the farm or enter some other life work'. Rather charmingly, the introductory page adds: 'It is evident that in young children the perceptive powers are most keenly alive. Everything is novel and interesting to them.' And also: 'Children love to do things.'

I would have taken a natural history GCSE in a heartbeat. At least, I think I would. At our secondary school there was a little plot of Portacabins and rabbit hutches and some kids studied – what was it called? Outdoor education? Anyway, a lot of us, more academic, more middle-class, were dickheads and called it *potato digging*. When I was sixteen, I'm not sure I would have gone for it. But when I was nine, I wouldn't have wanted to do anything else.

The natural history GCSE exists largely because of a campaign headed up by the formidable Mary Colwell, who for years worked in the BBC's natural history unit (on programmes including *Wildlife on One*, which ran for my entire childhood). It's needed, she believes, if the UK is to produce the naturalists and conservationists it's going to need. 'I would like a GCSE in Natural History to teach the skills of identifying, monitoring and recording the life around us,' she argues on her website. 'To know about migration and invasion of species. To understand how the seasons affect wildlife and how that is changing. I would like the history of studying natural history to be taught, from the earliest pioneers to Darwin, and the great natural scientists today. And nature literature, from Gilbert White to Robert Macfarlane. To understand how nature on TV and radio affects conservation. There are so many strands to be explored.'

I do want the next generation of young people – *all* the generations of young people – to know more about nature:

what it is, how it works, what we need to do (and what we need to *not* do) to help it be what it ought to be. I *really* like the idea of Gilbert White on the national curriculum.

I'm also aware, though, that the same cry goes up whenever someone 'exposes' a startling shortfall in our teenagers' knowledge of the workings of the world, whether it's to do with economics, politics, geography, food and agriculture, whatever: *get it on the curriculum*. Similarly, there's a certain sort of person who, on hearing of some historical event or person for the first time, reflexively demands: 'Why didn't I learn about this in school?' It could be anything, Mary Seacole, Peterloo, the Trail of Tears, the Bengal famine, the Malian empire, the Haitian Revolution, Muhammad Ali, anything – and then, from the other side, politically speaking, it's all 'get it *back* on the curriculum', because how is anyone supposed to learn about the Battle of Hastings, the repeal of the Corn Laws, the feudal system, El Alamein, the Lake Poets, the Civil War and Queen Victoria, if not in the classroom (or, they might concede, by visiting an ill-conceived civic statue)?

Not everything we need can be piped into the water supply.

I would like to live in a country of nature lovers and nature knowers. It's just – well, we don't live in a country of algebra enthusiasts and Shakespeare experts; conversation in our pubs and coffee shops doesn't revolve around oxbow lakes, erosion, osmosis and covalent bonds; most of us don't speak French; I very seldom throw the discus nowadays, and so on – it's not that school doesn't *work*, of course, but rather that I'm not sure it does quite enough. A lot hinges on how, over the next few generations, we connect with nature. Put it this way: if the survival of human societies as we know them depended on, let's say, familiarity with the novels of Dickens, we'd have had Danny chewing on the board-book edition of *Dombey and Son* before he could walk, and right about now we'd be walking Genevieve through her first reading lessons, *I-t*, it, *w-a-s*, was, *t-h-e*, the, *b-e-s-t*, best, *o-f*, of –

What scares me – and it might well be already here, for all I know – is the idea of an entire generation of sixteen-year-olds, 1.5 million kids or thereabouts, coming up to their GCSEs or SQCs and *needing* a GCSE in natural history.

Then there's what I'm going to call the St Snobs Question. Yes, we dutifully mouth, nature belongs to us all – but the reality is that *access* to nature, any kind of nature, is a privilege; familiarity with green spaces is a metric of social inequality. I'm not kidding myself, I was really lucky, with my primary school and its big field and handful of trees, with our 10 ft by 10 ft garden, with a half-decent park nearby, with suburbia. Loads of kids don't have these things. The heatwave we just had waved a big red flag at us in this respect: in towns and cities, districts with less greenery got hotter – people in less green districts, districts without street trees or shady parks or shrubs in precinct planters or living roofs or grassy verges, were, are, more at risk from extreme heat. The connection between privilege and access to greenery, access to nature, is – excuse a pun – a concrete one. Might the natural history GCSE be a posh kids' qualification? Of course, up to a point, *all* GCSEs are posh kids' qualifications – middle-class kids, kids whose parents went to university, are more likely to grow up with books in the house, get help with homework, have better access to decent IT and the internet, all that. But some subjects are more accessible than others; I worry that a GCSE in natural history might be a bit like a GCSE in lacrosse or skiing or listening to Radio 4 or barn conversion.

But I'm glad, anyway, that natural history is at least being thought about in the context of schools, and pre-schools, and the wider lives of our kids. A couple of weeks ago I took Genevieve for a visit to her new school, to meet the teachers, look around, explore. I marvelled at all the things that grown-ups always marvel at when they visit a primary school for the first time in a while: the tiny wee chairs! The tiny wee toilets! School dinners have changed since I was little – whither the spam fritter? – but the *smell* of school dinners hasn't. The noise of a room full of children hasn't. The monumental cast-iron radiators, slathered in a half-inch of cracked magnolia paint, were the same as they were at my old school. But never mind a sea urchin and a dead mole – each classroom here was like Fabre's study, decked out everywhere with leaves, rocks, bark and wood, berries, fossils, corals, feathers.

We went around looking at everything – touching everything. Danny will come here too, quite soon (Christ, in just, what, two years, how, *how*? – he was still in a baby carrier only a couple of chapters ago). Neither of them will look at all these things with surprise, with amazement, and I'm glad about that. I never want nature to be a novelty, because that's a scene from a dystopian TV series set in the near future, isn't it, the kids turning over a leaf or an eggshell or a feather in their hands, gazing at each other in wonder, what *was* this thing, what *lived* here? – I want nature to be an everyday thing, I want it to be familiar (familiar in two senses, firstly as something as close, as natural, as normal as family (our family is normal to us, if to no one else), and secondly as something that's *part* of our family, something we do together, something we've always done together).

Nature is always changing. That's the one thing we can know for sure about it, the one constant. Yesterday, today, tomorrow: we're always walking through a different wood. In the classroom I found a little box of tropical shells, the sort you buy in bags from seashell shops (I had loads when I was little). I fished around in it and found a cowrie. There was something satisfying in the sit of it in my hand, as there is with a cricket ball or a Guinness glass, but of course this wasn't just about shape and weight and texture, it wasn't that that made my throat lump up, just for a half-second, before I put it back down and went back to catching plastic fish in the water trough with Genevieve. Nature has changed and I've changed but there's something *between* me and nature, something that's never fallen away, that's the same. I still want to see it, to hold it, to know it. I see the same thing building, growing, when Genevieve picks a bluebell to take home to Mummy, when Daniel chases a pigeon across the park ('Dan, *Dan*,' I shout, like a man whose dog has slipped its lead).

It'll all change, everything will change. It's hard to see how it will all pan out. It's hard to see how anyone will make it all all right, whatever *all right* looks like. I don't know, and can hardly imagine, what part my children will play in it all – for me, for now, it's enough to know that they'll be in it somewhere, out there, learning, exploring, taking it in their hands.

FURTHER READING

This book, like most books, is a by-product of a lot of other people's books.

Many of the books I've written about here – many of the books that made me – can often be found in the Countryside section of second-hand bookshops, typically alongside books on houseplants or breeding beagles: Gerald Durrell's *The Amateur Naturalist* and the Reader's Digest *Book of British Birds* are certainly in this category.

You can learn more about Charles Waterton, Squire of Walton, in *Charles Waterton* by Julia Blackburn (Bodley Head, 1989); about Edward Wilson in two out-of-print books by George Seaver, *Edward Wilson: Naturalist and Friend* (John Murray, 1935) and *Edward Wilson: Nature Lover* (John Murray, 1937); about Sir Peter Scott in his memoir *The Eye of the Wind* (Hodder & Stoughton, 1961); about Mary Anning in *Jurassic Mary* by Patricia Pierce (The History Press, 2006) or *The Fossil Woman* by Tom Sharpe (Dovecote Press, 2021); about Philip and Edmund Gosse in Edmund's classic *Father and Son* (W. Heinemann, 1907) or his 1890 biography of his father, *The Naturalist of the Sea-shore* (W. Heinemann).

Emily Brady's wonderful 2011 paper 'The Ugly Truth: Negative Aesthetics and Environment' (*Royal Institute of Philosophy Supplement*, 69, 83–99) is, I imagine, the last word on ugliness in nature. Oscar Horta's 'Debunking the Idyllic View of Natural Processes: Population Dynamics and Suffering in the Wild' (*Télos*, 17, 2010, 73–88) is similarly comprehensive on life and suffering.

All of the children's books I write about are wonderful and should be sought out and bought by the bucketload. Steep the children in them (your children, someone else's children, all the children).

ACKNOWLEDGEMENTS

This book wouldn't exist – and even if it did would be far, far worse – without the expertise and encouragement of Clare Bullock, my editor at Icon. She was in on this thing from start to finish, and I'm more grateful than I can say for her know-how, support, wisdom and kindness.

Many people helped me out, in ways small or big, offering knowledge or counsel or a pint at the right time, while the book was getting written. Others contributed long before the book had even been conceived but made their mark on it anyway. Thank you to Dave Borthwick, James Lilford, Kay Peddle, George Harris, Hannah Lloyd-Hartley at the Dove Laboratory, Douglas Russell and Rosie Jones at the Natural History Museum, Gary Spaven, Helen Smith, Tony McGowan, Tom Shakespeare, 'Bill', Nick Acheson, Chris Foster, Patrick Galbraith, John Gallagher, Jules Howard, Heather Buttivant, Eamonn Griffin, Swaef, Gerry Cambridge and Martin Bradley.

Of course, I have to thank my mum and dad, my brother James and my Auntie Joan (my *up*family). My granny and grandad aren't around to thank any more, but they were part of it too – the childhood that made me, among other things, a massive nature dweeb, and that I'll never stop being grateful for.

Then there's Catherine and Genevieve and Daniel. My *down*family. Thank you for not minding about being put in a book (I hope you'll still not mind, kids, when you're old enough to read it). Thank you, always, for everything else.

INDEX